SHEFFIELD HALLAM UNIVERSITY
LEARNING CENTRE
CITY CAMPUS, POND
SHEFFIELD, S1 1
101 558 904
D1178911
£50-00
WITHDRAWN FROM STOCK

Extraction Methods for Environmental Analysis

Extraction Methods for Environmental Analysis

JOHN R. DEAN
University of Northumbria at Newcastle, UK

JOHN WILEY & SONS
Chichester · New York · Weinheim · Brisbane · Singapore · Toronto

National 01243 779777
International (+44) 1243 779777
e-mail (for orders and customer service enquiries): cs-books@wiley.co.uk
Visit our Home Page on http://www.wiley.co.uk
or http://www.wiley.com

Other Wiley Editorial Offices

John Wiley & Sons, Inc., 605 Third Avenue,
New York, NY 10158-0012, USA

WILEY-VCH Verlag GmbH, Pappelallee 3,
D-69469 Weinheim, Germany

Jacaranda Wiley Ltd, 33 Park Road, Milton,
Queensland 4064, Australia

John Wiley & Sons (Asia) Pte Ltd, Clementi Loop #02-01,
Jin Xing Distripark, Singapore 129809

John Wiley & Sons (Canada) Ltd, 22 Worcester Road,
Rexdale, Ontario M9W 1L1, Canada

Library of Congress Cataloging-in-Publication Data

Dean, John R.
Extraction methods for environmental analysis / John R. Dean
p. cm.
Includes bibliographical references and index.
ISBN 0–471–98287–3 (cloth : alk. paper)
1. Extraction (Chemistry) 2. Environmental chemistry—Technique.
I. Title
QD63.E88D43 1998 97–48744
628.5′01′543—dc21 CIP

British Library Cataloguing in Publication Data

A catalogue record for this book is available from the British Library

ISBN 0 471 98287 3

Typeset in 10/12 Times by Techset Composition Ltd, Salisbury, England
Printed and bound in Great Britain by Bookcraft (Bath) Ltd
This book is printed on acid-free paper responsibly manufactured from sustainable forestry, in which at least two trees are planted for each one used for paper production

To

Lynne, Samuel and Naomi

Contents

Preface

Pollution of the environment poses a threat to the health and wealth of every nation. Consequently it is essential to monitor the levels of organic pollutants in the environment. This book strives to highlight the traditional approaches of sample preparation for organic samples that have been, and continue to be, used whilst also considering modern alternatives. The reader is encouraged not only to use the book as a guide to the different approaches that are currently available, but also to consider what alternatives there may be just around the corner.

The book is broadly divided in to two areas: aqueous samples and solid samples. In the case of 'aqueous samples' the methods are based on approaches for preconcentrating the analytes from a large volume of water. In contrast, 'solid samples' involves methods for the extraction of analytes from solid or semi-solid samples. As the book is mainly concerned with the procedures for preconcentration/extraction, only a brief overview of chromatographic methods of analysis (Chapter 1) is provided. In addition, Chapter 1 covers introductory aspects for the sampling of aqueous and solid matrices, storage and preservation of samples, and quality assurance in environmental analysis.

Each area (aqueous or solid samples) is introduced to provide the essentials as to why it is necessary to monitor aqueous or solid samples. Aqueous samples are introduced by the use of a case study concerned with pesticides in the aquatic environment. In this way the reader is informed as to how pesticides are commonly introduced in to the aquatic environment, reasons as to why it is important to monitor the levels of pesticides, and the fate and behaviour of pesticides. Methods of preconcentration are then illustrated in subsequent chapters. The traditional approach for analyte preconcentration is based on liquid–liquid extraction, LLE. Chapter 3 outlines the theoretical and practical basis for effective LLE. As these traditional approaches invariably use large volumes of organic solvent it is necessary to 'preconcentrate' the extracts further. This is done using one of a variety of solvent evaporation methods (e.g. rotary evaporation, gas blow down, Kuderna–Danish evaporation and EVACS). Finally, the particular case of extraction of volatile organic compounds is illustrated via the technique of 'purge and trap'.

A modern alternative to LLE is solid phase extraction, SPE (Chapter 4). The principle of SPE is that analyte(s) from a large volume of an aqueous sample can be preferentially retained on a solid sorbent and then eluted with a small volume of organic solvent prior to analysis. The chapter covers the different types of SPE media available, whether in the form of a disk or, more commonly, as a cartridge, method of operation and solvent selection. In addition to preconcentration it is possible to utilise SPE for sample clean-up. The normal mode of operation for SPE is off-line, i.e. the analyte(s) are preconcentrated and then analysed separately. However recent trends, for specific purposes, have seen the introduction of on-line SPE systems directly coupled to either high performance liquid chromatography (HPLC) or gas chromatography (GC). Specific, selected examples are used to demonstrate both the off-line and on-line approaches.

The final chapter of this section is concerned with the latest development in aqueous sample preconcentration, solid-phase microextraction, SPME (Chapter 5). In this approach a silica-coated fibre is exposed to the sample for a predefined time (sampling), retracted into its protective casing and introduced into either the hot injector of a GC or eluted into an HPLC system using the mobile phase. The former is currently the most common approach. After theoretical and practical descriptions of SPME, selected applications of the use of SPME are reviewed for a range of analyte types (volatile organics, pesticides, phenols). Finally, the versatility of SPME is demonstrated by highlighting the novel approaches to which it has been applied.

Chapter 6 introduces the background for analysis of pollutants from solid samples. The chapter considers approaches for remediation of soil including containment, treatment and removal. The chapter then goes on to discuss some fundamental questions with regard to environmental analysis of polluted soils: how will you know that total recovery of the pollutant has occurred? What influence does the soil matrix have on the retention of the pollutants? And, which extraction techniques have approved methods? This last question then provides the appropriate technique information for discussion in subsequent chapters.

Extraction of organic pollutants from solid matrices is traditionally done using liquid–solid extraction, Chapter 7. Liquid–solid extraction can be sub-divided into approaches that utilise heat and those that do not. The use of heat is typified by Soxhlet extraction while the cold extraction methods by sonication or shake-flask. Experimental details for each type of extraction approach are presented as well as selected literature examples of the various liquid–solid extraction procedures.

Alternatives to the traditional liquid–solid extraction approaches are focused on instrumental methods, typically supercritical fluid extraction (Chapter 8), microwave-assisted extraction (Chapter 9) and accelerated solvent extraction (Chapter 10). Each of these extraction methods is discussed in terms of instrumentation and theoretical considerations. Environmental applications of each extraction technique are then highlighted with respect to a range of organic pollutants, for example, polycyclic aromatic hydrocarbons, polychlorinated biphenyls, phenols and pesticides. Specific emphasis is placed on describing particular features and/or applications of each technique.

The merits of each extraction technique for either aqueous or solid samples is summarised against a wide range of criteria to provide the reader with an easy-to-read comparison (Chapter 11). Finally, potential future developments for sample preparation are considered in the light of miniaturisation of scientific instruments.

John R. Dean
March 1998

1

Environmental Analysis

The analysis of environmental samples for organic pollutants is often a complicated procedure involving many steps. These steps culminate in the use of chromatographic separation coupled with a suitable detector. The effectiveness of the analysis does not depend, however, on the high cost of the chromatography equipment, though it is anticipated that, with due calibration and operating a suitable quality assurance scheme, it will provide accurate and precise data. The accuracy and precision of the data generated is not simply dependent on the chromatographic apparatus used but is based on a series of cumulative operations that have gone before. These include the sampling strategy, storage of the sample, sample pretreatment, sample extraction techniques to be utilised and, if necessary, extract clean-up and/or preconcentration. All these operations are probable sources of inaccuracy and imprecision that can inadvertantly be introduced into the entire analytical procedure. While preliminary information is given on all these aspects, it is the primary focus of this book to consider the apparatus, usage and application of extraction techniques for environmental analysis only.

1.1 INTRODUCTION

The major sources of environmental pollutants can be attributed to agriculture, electricity generation, derelict gas works, metalliferous mining and smelting, metallurgical industries, chemical and electronic industries, general urban and industrial sources, waste disposal, transport and other miscellaneous sources. Table 1.1 identifies some of the common pollutants and the environmental media in which they are found, adapted from Ref. 1 with particular focus on organic pollutants only. It is therefore not suprising to find that the UK, the European Community and the USA (to cite but three) have priority lists of pollutants that need to be routinely monitored. The UK priority or red list of pollutants is shown in Table 1.2. It is seen that while the UK list is not extensive it does contain a range of organic pollutants.

Table 1.1 Sources of organic pollutants found in the environment

Agricultural	Air	Pesticide aerosols
	Water	Pesticide spillages, run-off, soil particles; hydrocarbon (fuel) spillages
	Soil	Pesticides, persistent organics, e.g. DDT, lindane; fuel spillages (hydrocarbons)
Electricity generation	Air	Polycyclic aromatic hydrocarbons (PAHs) from coal
	Water	PAHs from ash
	Soil	Ash, coal dust
Derelict gas works sites	Air	Volatile organic compounds (VOCs)
	Water	PAHs, phenols
	Soil	Tars containing hydrocarbons, phenols, benzene, xylene, naphthalene and PAHs
Metallurgical industries	Air	VOCs
	Water	Solvents (VOCs) from metal cleaning
	Soil	Solvents
Chemical and electronic industries	Air	VOCs, numerous volatile compounds
	Water	Waste disposal, wide range of chemicals in effluents, solvents from microelectronics
	Soil	Particulate fallout from chimneys; sites of effluent and storage lagoons, loading and packaging areas; scrap and damaged electrical components, e.g. PAHs
General urban/industrial sources	Air	VOCs, aerosols (PAHs, PCBs, dioxins); fossil fuel consumption, e.g. PAHs; bonfires, e.g. PAHs, dioxins and furans
	Water	Wide range of effluents, PAHs from soot, waste oils, e.g. hydrocarbons, PAHs, detergents
	Soil	PAHs, polychlorinated biphenyls (PCBs), dioxins, hydrocarbons
Waste disposal	Air	Incineration–fumes, aerosols and particulates, e.g. dioxins and furans, PAHs; landfills, e.g. CH_4, VOCs; livestock farming wastes, e.g. CH_4; scrapyards–combustion of plastics, e.g. PAHs, dioxins and furans
	Water	Landfill leachates, e.g. PCBs
	Soil	Sewage sludge, e.g. PAHs, PCBs; scrapheaps, e.g. PAHs, PCBs; bonfires, e.g. PAHs; fallout from waste incinerators, e.g. furans, PCBs, PAHs; fly tipping of industrial wastes (wide range of substances); landfill leachate, e.g. PCBs
Transport	Air	Exhaust gases, aerosols and particulates, e.g. PAHs
	Water	Spillages of fuels; spillages of transported loads, e.g. hydrocarbons, pesticides and manufactured organic chemicals; wastes in transit, road and airport de-icers, e.g. ethylene glycol; deposition of fuel combustion products, e.g. PAHs

Table 1 (*continued*)

	Soil	Particulates, e.g. PAHs; wide range of soluble/ insoluble compounds at docks and marshalling yards and sidings, deposition of fuel combustion products, e.g. PAHs
Incidental sources	Water	Leakage from underground storage tanks, e.g. solvents, petrol products
	Soil	Preserved wood, e.g. pentachlorophenol, creosote
	All media	Warfare, e.g. fuels, explosives, ammunition, bullets, electrical components, poison gases, combustion products; industrial accidents, e.g. Bhopal, Seveso
Long-range atmospheric transport (deposition of transported pollutants)	Water and soil	Pesticides, PAHs; wind blown soil particles with adsorbed pesticides and pollutants

Table 1.2 UK priority or red list of environmental pollutants[2]

Mercury and its compounds	1,2-Dichloroethane
Cadmium and its compounds	Trichlorobenzene
Gamma-hexachlorohexane	Atrazine
DDT	Simazine
Pentachlorophenol	Tributyltin compounds
Hexachlorobenzene	Triphenyltin compounds
Hexachlorobutadiene	Trifluralin
Aldrin	Fenitrothion
Dieldrin	Azinphos-methyl
Endrin	Malathion
Polychlorinated biphenyls (PCBs)	Endosulfan
Dichlorvos	

1.2 SAMPLING STRATEGIES

The main objective in any sampling strategy is to obtain a representative portion of the sample. This requires a detailed plan of how to carry out the sampling. Therefore planning of the sampling strategy is an important part of the overall analytical procedure as the consequences of a poorly defined sampling strategy, as well as costing both time and money, could well lead to getting the wrong answer. Included below is a checklist of the necessary criteria for carrying out an effective sampling strategy for environmental analysis (adapted from reference 3):

- What are your data quality objectives? What will you do if these objectives are not met (i.e. resample or revise objectives)?

- Have arrangements been made to obtain samples from the sites? Have alternative plans been prepared in case not all sites can be sampled?
- Is specialised sampling equipment needed and/or available?
- Are the samplers experienced in the type of sampling required/available?
- Have all analytes been listed? Has the level of detection for each analyte been specified? Have methods been specified for each analyte? What sample sites are needed based on method and desired level of detection?
- List specific method quality assurance/quality control protocols required. Are there specific types of quality control samples? Does the instrument require optimisation of its operating parameters?
- What type of sampling approach will be used? Random, systematic, judgemental or a combination of these? Will the type of sampling meet your data quality objectives?
- What type of data analysis methods will be used? Chemometrics, control charts, hypothesis testing? Will the data analysis methods meet your data quality objectives? Is the sampling approach compatible with data analysis methods?
- How many samples are needed? How many sample sites are there? How many methods were specified? How many test samples are needed for each method? How many control site samples are needed? What types of quality control samples are needed? How many exploratory samples are needed? How many supplementary samples will be taken?

In addition, it is necessary to collect appropriate blank samples. Blank samples are matrices that have no measurable amount of the analyte of interest. The ideal blank will be collected from the same site as the samples but will be free of the pollutant. All conditions relating to collection of the blank sample, storage, pretreatment, extraction and analysis will be carried out as the actual samples. Once these questions can be answered it is then necessary to go and collect the samples.

1.2.1 SAMPLING WATER MATRICES

While natural water would appear to be homogeneous this is not in fact the case. Natural water is heterogeneous, both spatially and temporally, making it extremely difficult to obtain representative samples. Stratification within oceans, lakes and rivers is common with variations in flow, chemical composition and temperature. Variations with respect to time (temporal) can occur, for example because of heavy precipitation (snow, rain) and seasonal changes.

1.2.2 SAMPLING SOILS AND SLUDGES

In this situation sample hetereogeneity is assumed and the outlining of a suitable, statistical approach to sampling is essential. It may be that a particular highly contaminated part of a site is specifically targeted for analysis, not to be representative of the entire site but to provide the worst case scenario. Often, however, it is important, because of the heterogeneity, to select as large a test sample as is practical. This is because an extract of this sample will be more homogeneous, and it will provide more reproducible aliquots than a smaller portion of the sample.[3]

It is not the intention of this book to provide details of the actual sampling devices used and the procedures to follow. For this and related information the reader should consult more specialist texts.

1.3 STORAGE OF SAMPLES

It is probably unfortunate that most laboratories cannot analyse samples immediately upon receipt, so some form of sample storage is almost always required. The concern with the storage of samples is that losses can occur, due to adsorption to the storage vessel walls, or that potential contaminants can enter the sample, from desorption or leaching from the storage vessels. These problems can all lead to the analyst getting the wrong answer or at least an unexpected answer after the analysis has taken place. The goal therefore is to store samples for the shortest possible time interval between sampling and analysis. Indeed in some instances where analytes are known to be unstable or volatile it may be necessary to perform the analysis immediately upon receipt or not at all.

For environmental type samples, e.g. water, soils or sludges, it is common to find that samples are stored in the refrigerator at 4 °C. Although if the analyte within the matrix is known to degrade, it may be more appropriate to place the sample in the freezer at −18 °C. Storage under these conditions reduces most enzymatic and oxidative reactions. If storage is required it should be noted how long the sample has been stored and under what conditions storage has been done.

The nature and type of storage vessel is also important. For example if it is known that the analyte is light sensitive it is then essential that the sample is stored in a brown glass container to prevent photochemical degradation. For volatile species it is also desirable that the sample is stored in a well-sealed container. In most cases, the use of glass containers is recommended as there is little opportunity for contamination to result as a consequence of the vessel itself. It is also important that the appropriate sized container is used. It is better to completely fill the storage container rather than leave a significant headspace above the sample. This acts to reduce any oxidation that may occur. In addition to glass containers, polyethylene or poly(tetrafluoroethylene) (PTFE) containers are appropriate to use for solid samples. Plastic containers are not recommended for aqueous samples as plasticisers are prone to leach from the vessels which can cause problems at later stages of the analysis, e.g. phthalates.

Whatever method of storage is chosen it is desirable to perform experiments to identify that the analyte of interest does not undergo any chemical or microbial degradation and that contamination is kept to a minimum risk.

A recent review[4] highlighted the problems associated with the storage and preservation of polar pesticides in water samples. The authors identified that in some instances the degradation products of pesticides are more stable than the parent compounds so that emphasis, in terms of the analysis, should also be aimed at the degradation products. For example Chiron *et al.*[5] found that some carbamate pesticides (methiocarb sulfone, methiocarb sulfoxide and 3-ketocarbofuran) are stable in water samples for up to 20 days whereas carbaryl losses can be as high as 90% in one day.[6] A summary of the recommended standard preservation techniques for pesticides is presented in Table 1.3 (adapted from reference 4). An alternative to preservation of the aqueous samples at 4 °C is the use of solid phase extraction (SPE) disks or cartridges (see Chapter 4).[4] In this case the aqueous sample is passed through the SPE media and retained until required for analysis. At that point the analytes are eluted and the chromatographic assay completed.

1.4 BRIEF INTRODUCTION TO PRACTICAL CHROMATOGRAPHIC ANALYSIS

It is not the intention of this section to give a complete and detailed study of chromatographic analysis but to provide a general overview of the types of separation frequently used in environmental organic analysis. The most common two approaches for separation of an analyte from other compounds in the sample extract are gas chromatography (GC) and high performance liquid chromatography (HPLC). The essential difference between the two techniques is the nature of the partitioning process. In GC the analyte is partitioned between a stationary phase and a gaseous phase, whereas in HPLC the partitioning process occurs between a stationary phase and a liquid phase. Separation is therefore achieved in both cases by the affinity of the analyte of interest with the stationary phase; the higher the affinity, the more the analyte is retained by the column. The choice of which technique is employed is largely dependent upon the analyte of interest. For example if the analyte of interest is thermally labile, does not volatilise at temperatures up to 250 °C and is strongly polar then GC is not the technique for it, however, HPLC can then be used (and vice versa).

Separation in GC is based on the vapour pressures of volatilised compounds and their affinity for the liquid stationary phase, which coats a solid support, as they pass down the column in a carrier gas. The practice of GC can be divided into two broad categories, packed and capillary column based. For the purpose of this discussion only capillary column GC will be considered. A gas chromatograph (Figure 1.1) consists of a column, typically 15–30 m long with an internal diameter

Table 1.3 Container preservation techniques[a] for pesticides in water

Pesticides	Container	Preservation	Maximum holding time
Organochlorine	Glass	1 ml of a 10 mg ml^{-1} $HgCl_2$ or 1 g l^{-1} ascorbic acid or adding of extraction solvent	7 days, 40 days after extraction
Organophosphorus	Glass	1 ml of a 10 mg ml^{-1} $HgCl_2$ or adding of extraction solvent	14 days, 28 days after extraction
Chlorinated herbicides	Glass	Refrigeration, sealed, add HCl to $pH < 2$	14 days
		Sodium thiosulfate	7 days, 28–40 days after extraction
		1 ml of 10 mg ml^{-1} $HgCl_2$	7 days
Phenolic compounds	Glass	Refrigerate, add H_2SO_4 to $pH < 2$	28 days
Polar pesticides	Glass	1 ml of 10 mg ml^{-1} $HgCl_2$	28 days

[a] As recommended by different agencies (EPA and ISO). Storage temperature is usually at 4 °C. Sampling of 1 litre of water, do not prerinse bottles with sample before collection. Seal bottles and shake vigorously during 1 min. Refrigerate samples until extracted. Protect from light.

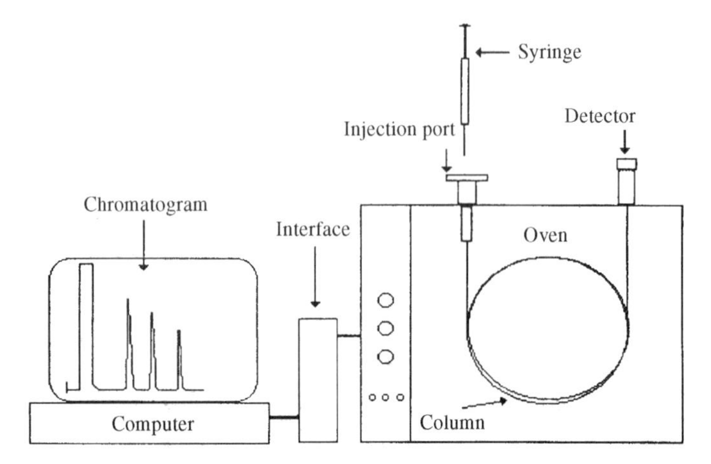

Figure 1.1 Schematic diagram of a gas chromatograph. Reproduced by permission of Mr E Ludkin, University of Northumbria at Newcastle

of 0.1–0.3 mm. The range of column types available from manufacturers is considerable, however, some common types are frequently encountered, e.g. DB-5. The DB-5 is a low polarity column in which the stationary phase, 5% phenyl and 95% methylsilicone, is chemically bonded onto silica. The thickness of the stationary phase is of the order of 0.25 μm. The column, through which the carrier gas passes, is placed in a temperature-controlled oven. The carrier gas (e.g. nitrogen) is supplied from a cylinder. Temperature programming allows a low initial temperature to be maintained to allow the separation of high boiling point analytes, this is then followed by a stepwise or linear temperature increase to separate analytes with lower boiling points. Typical column temperature changes can range from 50 to 250 °C. In order to introduce the sample requires an injector, of which there are several types. The aim of using an injector is to introduce a small but representative portion of the sample onto the column without overloading. Sample is introduced into the injector by means of a hyperdermic syringe. Two common approaches are applied. The first allows a larger sample volume (μl) to be introduced into the injection port and then 'splits' or divides the sample, the split/splitless injector. In this case a large volume of sample is introduced into the heated injection port where it is instantly vaporised but only a small proportion is introduced into the column, the rest being vented to waste. The ratio of the split flow to the column flow is called the split ratio and can be of the order of 50 : 1 or 100 : 1. The other type of injector introduces a smaller volume (nl) of sample directly onto the column, the cold on-column injector. In this case a syringe fitted with a very long thin needle is required to introduce the sample directly onto the column. The range of detectors available range from the universal (flame ionization detector) to the specific (electron capture, thermionic, flame photometric, atomic emission detectors). For example the electron capture detector is specific for halogen-containing compounds, e.g. organochlorine pesticides. In addition, GC has been coupled to mass spectrometry to provide a highly sensitive detector which also provides information on the molecular structure of the analyte. For a more detailed discussion of both the theoretical aspects and practical details the reader is referred to the bibliography section of this chapter.

High performance liquid chromatography can also be classified into two broad categories, normal phase and reversed phase. For the purpose of this discussion the most popular of these categories, reversed phase HPLC will only be considered. In reversed phase HPLC the stationary phase is non-polar and the mobile phase is polar. A high performance liquid chromatograph (Figure 1.2) consists of a column, typically 25 cm long with an internal diameter of 4.6 mm packed with a suitable stationary phase (octadecylsilyl or ODS which consists of a C18 hydrocarbon chain bonded to silica particles of 5–10 μm diameter) through which is passed a mobile phase. The mobile phase is a water–solvent system, with typical solvents being methanol or acetonitrile, which is pumped using a reciprocating or piston pump at a flow rate of 1 ml min^{-1}. The column is normally located in a column oven which is maintained at approximately 30 °C. Samples (10–20 μl) are injected, via a fixed volume loop connected to a six-port injection valve, onto the column and after separation is detected. The most common detector used for HPLC is the ultraviolet-visible spectrometer (available as a single wavelength unit or with a photodiode array that allows multiple wavelength detection), although a range of more

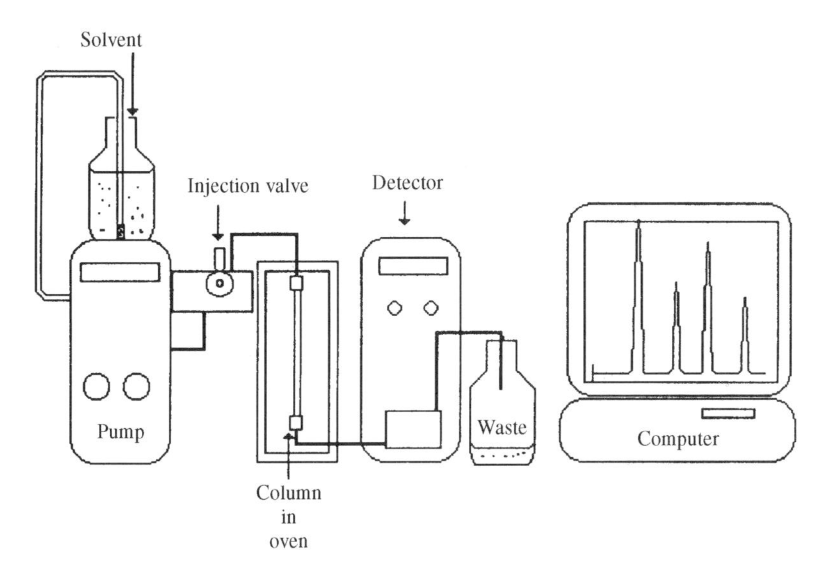

Figure 1.2 Schematic diagram of a high performance liquid chromatograph. Reproduced by permission of Mr E Ludkin, University of Northumbria at Newcastle

specialised detectors are also available, e.g. fluorescence, electrochemical, refractive index, light-scattering or chemiluminescence. Recently the introduction of low-cost benchtop liquid chromatograph–mass spectrometers has made the use of this universal detector with the capability of mass spectral interpretation of unknowns more readily available. The HPLC system can be operated in either isocratic mode, i.e. the same mobile phase composition throughout the chromatographic run or by gradient elution, i.e. the mobile phase composition varies with respect to run time. The choice of gradient or isocratic operation depends largely on the number of analytes to be separated and the speed with which the separation is required to be achieved. For a more detailed discussion of both the theoretical aspects and practical details the reader is referred to the bibliography section of this chapter.

1.5 QUALITY ASSURANCE IN ENVIRONMENTAL ANALYSIS

Quality assurance is about getting the correct result. However, while attention is primarily given to the analysis in the laboratory, environmental analysis and monitoring involves many more sampling handling steps, e.g. sample collection, treatment and storage, prior to laboratory analysis. It is likely that the variation in the final measurement is more influenced by the work external to the analytical laboratory than that within the laboratory.

In order to achieve good accuracy (the closeness of a measured value to the 'true' value) and precision (the measure of the degree of agreement between replicate analyses of a sample), in the laboratory at least, it is desirable that a good quality assurance scheme is operating. The main objectives of such a scheme are as follows:

- To select and validate appropriate methods of analysis.
- To maintain and upgrade analytical instruments.
- To ensure good record keeping of methods and results.
- To ensure quality of data produced.
- To maintain a high quality of laboratory performance.

The quality of data produced in the laboratory is controlled by the use of a good quality control procedure. In implementing a good quality control programme it is necessary to take into account the following:

- Certification of operator competence. This is intended to assess whether a particular operator can carry out sample and standard manipulations, operate the instrument in question and obtain data of appropriate quality. The definition of suitable data quality is open to interpretation but may be assessed in terms of replicate analyses of a check sample.

- Recovery of known additions. Samples are spiked with known concentrations of the same analyte and their recoveries noted. This approach will also allow the operator to determine whether any matrix effects are interfering with the analysis.
- Analysis of certified reference materials. By definition a certified reference material (CRM) is a substance for which one or more analytes have certified values, produced by a technically valid procedure, accompanied with a traceable certificate and issued by a certifying body. Common examples of certifiying bodies are the National Institute for Standards and Technology (NIST) based in Washington DC, USA; the Community Bureau of Reference (known as BCR), Brussels, Belgium; and, the Laboratory of the Government Chemist (LGC), London, UK.
- Analysis of reagent blanks. Analyse reagents whenever the batch is changed or a new reagent introduced. Introduce a minimum number of reagent blanks (typically 5% of the sample load); this allows reagent purity to be assessed and if necessary controlled and also acts to assess the overall procedural blank.
- Calibration with standards. A minimum number of standards should be used to generate the analytical curve, e.g. 6 or 7. Daily verification of the working curve should also be done using one or more standards within the linear working range.
- Analysis of duplicates. This allows the precision of the method to be reported.
- Maintenance of control charts. Various types of control charts can be maintained for standards, reagent blanks and replicate analytes. The purpose of each type of chart is to assess the longer term performance of the laboratory, instrument, operator or procedure, based on a statistical approach.

REFERENCES

1. B.J. Alloway and D.C. Ayres, *Chemical Principles of Environmental Pollution*, Blackie Academic and Professional, Glasgow (1993).
2. *UK Priority or Red List of Environmental Pollutants*, Department of the Environment, HMSO, London (1988).
3. L.H. Keith, *Environmental Sampling and Analysis: A Practical Guide*, Lewis Publishers, Inc., Chelsea, MI, USA (1991).
4. D. Barcelo and M.F. Alpendurada, *Chromatographia*, **42** (1996) 704.
5. S. Chiron, A. Fernandez-Alba and D. Barcelo, *Environ. Sci. Technol.*, **27** (1993) 2352.
6. M.P. Marcos, S. Chiron, J. Gascon, B.D. Hammock and Barcelo, D., *Anal. Chim. Acta*, **311** (1995) 319.

BIBLIOGRAPHY

A. Braithwaite and F.J. Smith (1990) *Chromatographic Methods* (4th edition), Chapman and Hall, London.

I.A. Fowlis (1995) *Gas Chromatography* (2nd edition), Analytical Chemistry by Open Learning, John Wiley & Sons Ltd, Chichester.

P.J. Schoenmakers (1986) *Optimization of Chromatographic Selectivity. A Guide to Method Development*, Elsevier Science Publishers BV, Amsterdam.

L.R. Synder and J.J. Kirkland (1979) *Introduction to Modern Liquid Chromatography* (2nd edition) John Wiley & Sons Ltd, New York.

L.R. Synder, J.J. Kirkland and J.L. Glajch (1997) *Practical HPLC Method Development* (2nd edition), John Wiley & Sons Ltd, New York.

I

AQUEOUS SAMPLES

2

Aqueous Sample Preparation

Aqueous sample preparation is an approach whereby a pollutant is determined, usually at low concentration, from a large volume of sample. The common approaches used are therefore based on methods of preconcentration. Details of the traditional approaches and newer methods are presented in subsequent chapters. The purpose of this chapter is to establish why it is necessary to monitor for organic pollutants in aqueous samples.

Aqueous samples can be subdivided into natural waters and waste waters. Natural waters consists of ground, surface and drinking waters. Ground water is water that occupies the subterranean permeable layers. The main source of ground water is through the percolation of rain water. The composition of ground water can vary depending upon geological and pedological properties. Surface waters can either be flowing (e.g. rivers) or still (e.g. lakes). Drinking water can be obtained from either ground or surface waters, and it is the quality of the water that makes it fit for human consumption. Finally, the term waste water is applied to water which has had its composition and properties changed by human activity. The waste water, e.g. from towns, factories and agriculture, can be a source of contamination of ground and surface waters. In order to protect the environment and to supply suitable potable water requires the introduction of suitable monitoring programmes. Using pesticides as an example, the sources, potential for pollution and the environmental fate of pesticides in natural waters is highlighted. However, the reader should bear in mind that pesticides are but one class of organic pollutants, albeit a large class of compounds, that are found in water.

2.1 ENVIRONMENTAL CASE STUDY: PESTICIDES

Pesticides are used for a variety of applications, e.g. agriculture and industry, by local authorities and in domestic situations. The pesticides used include the following:[1]

- weed killers
- mould killing agents
- insecticides
- fungicides
- slug killers
- rat poison
- soil sterilants
- antifouling paints
- food storage protectors
- industrial and domestic pest control products
- plant growth regulators
- products to prevent lichen, moss or fungal growth on buildings
- desiccants
- similarly acting chemicals used for the preservation of animal products used e.g. in the tanning and wool processing industries
- wood preservatives
- medicines which act in a pesticidal manner such as sheep dip chemicals, sea louse control for salmon and louse control for humans; and
- bird repellants.

The contamination of pesticides in surface water and ground water can be attributed to several sources. It has been reported[1] that the main pesticide pollution of surface waters from a point source, i.e. a specific, identifiable situation that leads to the release of a pesticide in to the aquatic environment, occurs due to:

- effluent from pesticide pollution
- effluent from industrial pesticide use, e.g. carpet manufacturing, leather and wool processing
- direct introduction of pesticides, e.g. for weed control in rivers, insect control at sewage treatment works and pest control in fish farms
- spillage of pesticides
- effluent from domestic sewage treatment.

In contrast, diffuse input (i.e. non-specific sources) into surface water can arise from the following:[1]

- spray drift from approved application of a pesticide to land
- accidental overspray of water during application
- long distance drift of the more volatile pesticides and of wet and dry aerosols formed during application
- atmospheric deposition and rainfall
- leaching from disposal sites
- leaching from soakaways, e.g. used for the disposal of sheep dip or horticultural bulb washings

- surface run-off and subsurface drain flow after a rainfall event from either agricultural or non-agricultural use
- inflow of contaminated ground water.

The situation for pesticide pollution of ground water is simpler. No direct application of pesticide is allowed to ground water, however, contamination can occur if, for example disused quarries or mines have been used for disposal.[1] This unexpected input to ground water can be a major problem due to its unexpected source of input.

Ground water can, however, be contaminated indirectly from a number of sources:[1]

- leaching from inappropriate use of agricultural soakaways
- leaching of pesticides through the soil and subsoil following approved application of pesticides to land
- leaching from disposal sites and road- and rail-side soakaways
- recharge of aquifers by river water contaminated with pesticides
- overspray around wells and boreholes.

Control of input to surface and ground water must therefore be limited by legislation and by constant monitoring of the levels of pesticides present. For monitoring to be effective requires that a suitable sampling frequency is established at selected, relevant sites followed by the appropriate analytical techniques to reach the desired level. Inherent in any analytical monitoring scheme is the adherence to a suitable quality assurance scheme (see Section 1.5). In order to protect the quality of potable and surface water in Europe a priority list of pesticides has been compiled, also called 'red' and/or 'black lists'.[2] The EEC Directive on the Quality of Water Intended for Human Consumption sets a maximum admissible concentration (MAC) of $0.1\,\mu g\,l^{-1}$ per individual pesticide.[3] This level is irrespective of the known toxicity of the pesticide concerned. For a few selected pesticides, even more stringent limits are in operation. Cypermethrin has an Environmental Quality Standard (EQS) of $1\,ng\,l^{-1}$ in natural water while for tributyltin it is $2\,ng\,l^{-1}$ in marine water. It is desirable that techniques are available to measure to one-tenth of the EQSs for these pesticides. A recent survey by the Drinking Water Inspectorate for England and Wales reported that the total number of individual pesticide determinations carried out by water supply and water service companies in 1993 was 1006458.[4] Of these determinations, 8.5% were attributable to the following triazine herbicides: atrazine, prometryn, propazine, simazine, terbutryne and trietazine. The most popular triazine monitored was atrazine (with 37647 determinations) closely followed by simazine (with 36130 determinations). The number of determinations contravening the MAC was 18.8% and 13.0% for atrazine and simazine, respectively. In the report,[4] the maximum detected concentration of atrazine was $5.57\,\mu g\,l^{-1}$; but only a single sample was reported containing this concentration.

In the USA, a joint project between the Environmental Protection Agencies (EPAs) Office of Drinking Water (ODW) and the Office of Pesticide Programs carried out a National Pesticide Survey.[5,6] The results were released between 1990 (Phase I)[5] and 1992 (Phase II).[6] The result is one of the most comprehensive lists used to conduct a monitoring programme on pesticides. The list arranges the pesticides according to their method of analysis. For example the triazines (ametryn, atrazine, prometone, prometryn, propazine, simazine, simetryn and terbutryne) are included in Method 1 (EPA method 507: dichloromethane extraction followed by gas chromatography (GC) with nitrogen phosphorus detection (NPD)).

It was reported[1] that the extent of pollution depends on many sources and in particular the following:

- the type of surface to which the pesticide is applied
- the crop and soil type
- the weather
- the nature of application
- the application rate
- the equipment used to apply and contain the pesticide
- the physical and chemical characteristics of the pesticide or its formulation.

A list of the 20 most used active ingredients is shown in Table 2.1.[1] The table contains information on ten herbicides, eight fungicides, one growth regulator and one desiccant. The desiccant, sulfuric acid, which is applied to potato crops and bulbs, accounted for 32% of the total weight of active ingredients applied to all crops in Great Britain in 1993.[1]

2.1.1 ENVIRONMENTAL FATE AND BEHAVIOUR OF PESTICIDES

The production of a pesticide is quite unlike most other chemicals, in as much as it is designed to be released deliberately into the environment to control disease and pests. In order to assess whether the release of a pesticide will have other environmental consequences it is necessary to evaluate potential risks, so that steps can be taken to minimise the environmental impact. Assessing this environmental impact requires the answers to several questions:[1]

- How will the pesticide be used?
- How likely is the pesticide and/or its transformation products to move from the area of application to other environmental compartments such as water, soil, air and biota?
- How long will this dispersion through the environment take?
- What will the resulting concentrations be in each environmental compartment?
- How persistent is the pesticide and/or its transformation products?

Table 2.1 Estimated amount (tonnes) of the 20 active ingredients, used most by weight, on all agricultural and horticultural crops grown in Great Britain in 1993[1]

	Active ingredient	Weight (tonnes)
1	Sulfuric acid	10167
2	Isoproturon	2809
3	Chlormequat	2416
4	Mancozeb	1361
5	Chlorothalonil	984
6	Sulfur	740
7	MCPA	730
8	Mecoprop	694
9	Fenpropimorph	650
10	Mecoprop-P	600
11	Chlorotoluron	580
12	Maneb	569
13	Pendimethalin	516
14	Trifluralin	382
15	Glyphosate	312
16	Carbendazim	297
17	Fenpropidin	295
18	Tri-allate	262
19	Metamitron	247
20	Tridemorph	238

- Will the use of the pesticide give rise to concentrations in the environment of toxicological concern?

In order to assess these questions requires additional data, that must be generated from a range of laboratory and field studies. The use of predictive models may also be required. The following properties were identified as important:[1]

Physico-chemical properties which influence environmental mobility

- chemical structure
- molecular weight
- melting/boiling point
- physical state at 20–25 °C
- vapour pressure at 20–25 °C
- solubility in water at 20–25 °C
- Henry's law constant
- partition coefficient between water and n-octanol (K_{ow})
- coefficient between soil/sediment organic carbon and water (K_{oc})
- acid/base dissociation constant (pK_a).

Degradation characteristics

Upon release into the environment a pesticide is subject to various processes that can lead to degradation:

- hydrolysis (chemical breakdown by water)
- photodegradation (breakdown by sunlight)
- reduction or oxidation
- microbial degradation.

It is important to identify the degradation products as their environmental fate will also need to be considered. In addition to laboratory studies, field trials are essential so that different climatic and soil characteristics can be investigated.

Mobility in soil

The mobility of the pesticide, after application in the field, is important in determining its potential to enter surface or ground waters. The mobility of the pesticide in soil is strongly dependent upon the nature of the soil. For example a pesticide with a high hydrophobicity will be potentially strongly absorbed by soil containing a high humus content and hence should be relatively immobile; this situation is the reverse in a sandy soil. So adsorption of a pesticide to soil is strongly dependent upon the nature and type of soil to which it has been applied. In addition, and also of importance, is the pesticide's ability to desorb from the soil. Desorption is controlled by the physico-chemical properties of the pesticide itself. Thus the mobility of pesticides through soil is dependent upon a combination of adsortion and desorption properties.

Toxicity to aquatic organisms

Upon entering the water course, the fate of the pesticide must also be determined. As in soil, the potential for degradation must be studied and an assessment made of the toxicity to aquatic life. The nature of the toxicity to aquatic organisms is dependent upon uptake, metabolism, transport within the organism and the ability of the organism to excrete the pesticide/degradation product. Bioaccumulation is assessed on the basis of the physico-chemical property of the n-octanol-water partition coefficient, K_{ow}. Thus pesticides with a $\log K_{ow} < 3$ are unlikely to bioaccumulate; pesticides with $\log K_{ow} > 3$ but < 6 are likely to bioaccumulate. Pesticides with a $\log K_{ow} > 6$ are unlikely to bioaccumulate since they are either insufficiently soluble or they are too large to enter the cellular structure of an organism.

The information presented in this chapter is by way of a brief introduction as to why measures are needed to monitor and control the level of pesticides in natural waters. Subsequent chapters deal with the methods that have been historically used and methods that are becoming more popular.

REFERENCES

1. Pesticides in water: Report of 'The Working Party on the Incidence of Pesticides in Water', Department of the Environment, HMSO, May 1996
2. D. Barcelo, *J. Chromatogr.*, **643** (1993) 117.
3. M. Fielding, D. Barcelo, A. Helweg, S. Galassi, L. Torstenson, P. van Zoonen, R. Wolter and G. Angeletti, in *Pesticides in Ground and Drinking Water* (Water Pollution Research Report, 27), Commission of the European Communities, Brussels, 1992, 1–136.
4. Drinking Water Inspectorate, a report by the Chief Inspector, Drinking Water 1993, HMSO, London, 1994.
5. US Environmental Protection Agency, *National Survey of Pesticides in Drinking Water Wells, Phase I Report, EPA PB-91-125765*, National Technical Information Services, Springfield, VA, 1990.
6. US Environmental Protection Agency, *National Survey of Pesticides in Drinking Water Wells, Phase II Report, EPA 570/9-91-020*, National Technical Information Services, Springfield, VA, 1992.

3

Classical Approaches for the Extraction of Analytes from Aqueous Samples

The most commonly used approach for the extraction of analytes from aqueous samples is liquid–liquid extraction and this is the major focus of this chapter. However, the chapter also contains a brief description of the purge and trap technique which is used for volatile organic compounds.

3.1 LIQUID–LIQUID EXTRACTION

The principal of liquid–liquid extraction is that the sample is distributed or partitioned between two immiscible solvents in which the analyte and matrix have different solubilities. The main advantages of this approach is the wide availability of pure, solvents and the use of low-cost apparatus.

3.1.1 THEORY OF LIQUID–LIQUID EXTRACTION

Two terms are used to describe the distribution of an analyte between two immiscible solvents: distribution coefficient and the distribution ratio.

The distribution coefficient is an equilibrium constant that describes the distribution of an analyte, A, between two immiscible solvents, e.g. an aqueous and an organic phase. For example an equilibrium can be obtained by shaking the aqueous phase containing the analyte, A, with an organic phase, such as hexane. This process can be written as an equation:

$$A(aq) \Longleftrightarrow A(org) \quad (1)$$

where (aq) and (org) are the aqueous and organic phases, respectively. The ratio of the activities of A in the two solvents is constant and can be represented by:

$$K_d = \{A\}_{org}/\{A\}_{aq} \tag{2}$$

where K_d is the distribution coefficient. While the numerical value of K_d provides a useful constant value, at a particular temperature, the activity coefficients are neither known nor easily measured.[1] A more useful expression is the fraction of analyte extracted (E), often expressed as a percentage:[2]

$$E = C_o V_o/(C_o V_o + C_{aq} V_{aq}) \tag{3}$$

or

$$E = K_d V/(1 + K_d V) \tag{4}$$

where C_o and C_{aq} are the concentrations of the analyte in the organic phase and aqueous phases, respectively; V_o and V_{aq} are the volumes of the organic and aqueous phases, respectively; and, V is the phase ratio V_o/V_{aq}.

For one step liquid–liquid extractions, K_d must be large, i.e. 10, for quantitative recovery ($>99\%$) of the analyte in one of the phases, e.g. the organic solvent.[2] This is a consequence of the phase ratio, V, which must be maintained within a practical range of values: $0.1 < V < 10$ (equation (4)). Typically, two or three repeat extractions are required with fresh organic solvent to achieve quantitative recoveries. Equation (5) is used to determine the amount of analyte extracted after successive multiple extractions:

$$E = 1 - [1/(1 + K_d V)]^n \tag{5}$$

where n = number of extractions. For example if the volume of the two phases is equal ($V = 1$) and $K_d = 3$ for an analyte, then four extractions ($n = 4$) would be required to achieve $>99\%$ recovery.

It can be the situation that the actual chemical form of the analyte in the aqueous and organic phases is not known, e.g. a variation in pH would have a significant effect on a weak acid or base. In this case the distribution ratio, D, is used:

$$D = \frac{\text{concentration of A in all chemical forms in the organic phase}}{\text{concentration of A in all chemical forms in the aqueous phase}}$$

(Note: For simple systems, when no chemical dissociation occurs, the distribution ratio is identical to the distribution coefficient.)

3.1.2 SOLVENT EXTRACTION

Two common approaches are possible. In the first approach the extraction is carried out discontinuously where equilibrium is established between two immiscible phases, or the second approach, continuous extraction. In the case of the latter, equilibrium may not be reached. The selectivity and efficiency of the extraction process is critically governed by the choice of the two immiscible solvents. Using aqueous and organic (e.g. dichloromethane, chloroform, ethylene acetate, toluene, etc.) solvent pairs of solvents, the more hydrophobic analytes prefer the organic solvent while the more hydrophilic compounds prefer the aqueous phase. The more desirable approach is quite often reflected in the nature of the target analyte. For example if the method of separation to be used is reversed-phase high performance liquid chromatography (HPLC), then the target analyte is best isolated in the aqueous phase. In this situation the target analyte can then be injected directly into the HPLC system. (Note: The target analyte may well require preconcentration, e.g. solid phase extraction (see Chapter 4), to achieve the appropriate level of sensitivity.) In contrast, if the target analyte is to be analysed by gas chromatography it is best isolated in the organic solvent. In addition, isolation of the target analyte in the organic phase allows solvent evaporation to be employed (see Section 3.1.3) thus allowing concentration of the target analyte.

The equilibrium process can be influenced by several factors that include adjustment of pH to prevent ionisation of acids or bases, by formation of ion-pairs with ionisable analytes, by formation of hydrophobic complexes with metal ions, or by adding neutral salts to the aqueous phase to reduce the solubility of the analyte (salting out).

In discontinuous extraction the most common approach uses a separating funnel (Figure 3.1). In this case the aqueous sample (1 l, at a specified pH) is introduced into a large separating funnel (2 l capacity with Teflon stopcock) and then 60 ml of a suitable organic solvent, e.g. dichloromethane is added. Seal and then shake the separating funnel, either manually or mechanically, vigorously for 1–2 min. This shaking process allows thorough interspersion between the two immiscible solvents, thereby maximising the contact between the two solvent phases and hence assisting mass transfer, thus allowing efficient partitioning to occur. It is necessary to periodically vent the excess pressure generated during the shaking process. After a suitable resting period (10 min) the organic solvent is collected in a flask. Fresh organic solvent is then added to the separating funnel and the process repeated again. This should be done at least three times in total. The three organic extracts should be combined, ready for concentration (see Section 3.1.3).

In some cases the kinetics of the extraction can be slow, such that, the equilibrium of the analyte between the aqueous and organic phases is poor, i.e. K_d is very small, or if the sample is large then continuous liquid–liquid extraction can be used. In this situation, fresh organic solvent is boiled, condensed and allowed to percolate repetitively through the analyte containing the aqueous sample. Two

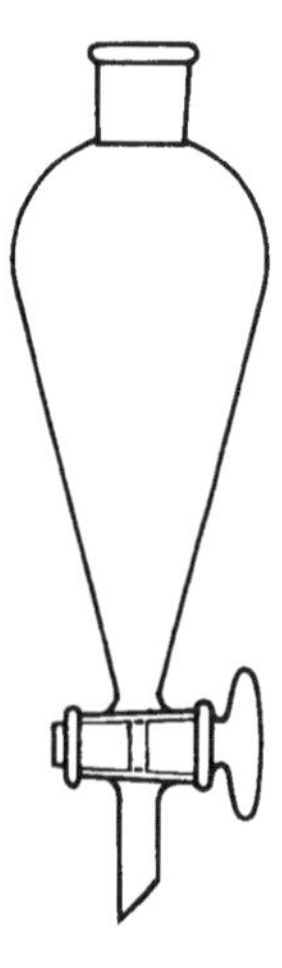

Figure 3.1 Separating funnel

common versions of continuous liquid extractors are available, using either lighter-than or heavier-than water organic solvents (Figure 3.2). Extractions usually take several hours, but do provide concentration factors up to 10^5. Obviously several systems can be operated unattended and in series, allowing multiple samples to be extracted. Typically, a 1 l sample, pH adjusted if necessary, is added to the continuous extractor. Then organic solvent, e.g. dichloromethane (in the case of a system in which the solvent has a greater density than the sample), of volume 300–500 ml is added to the distilling flask together with several boiling chips. The solvent is then boiled, in this case with a water bath, and the extraction process allowed to occur for between 18–24 hours. After completion of the extraction process, and sufficient cooling time, the boiling flask is detached and solvent evaporation can then occur (see Section 3.1.3).

Unfortunately as with most things, liquid–liquid extraction can suffer from some problems. The formation of emulsions is troublesome, particularly for samples that contain surfactants or fatty materials, if they cannot be broken up with, for example a centrifuge, filtration through a glass wool plug, refrigeration, salting out or the addition of a small amount of a different organic solvent. In addition, the rate of extraction may be different for the same analyte depending on the nature of the sample matrix. Obviously, as in all analysis the problem of controlling the level of contamination is crucial. It is essential to use the highest purity solvents (as any subsequent concentration may also concentrate impurity as well as the analyte of interest) and to wash all associated glassware thoroughly. As well as contamination, care should also be exercised to minimise analyte losses due to adsorption on glass containers.

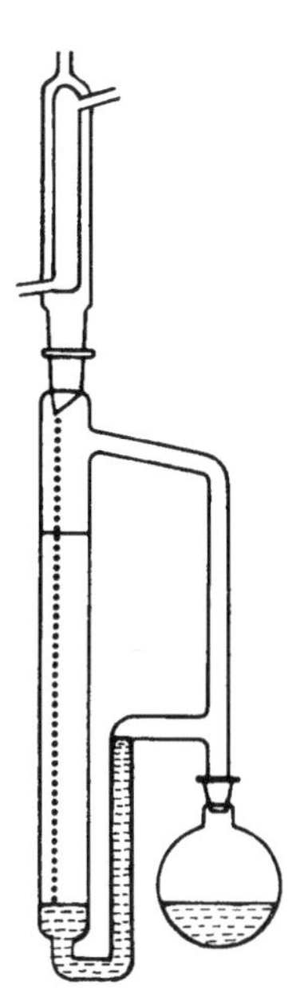

Figure 3.2 Continuous liquid–liquid extraction (organic solvent heavier than water)

3.1.3 SOLVENT EVAPORATION METHODS

The most common approaches for solvent evaporation are rotary evaporation, Kuderna–Danish evaporative concentration, automated **eva**porative **c**oncentration **s**ystem (EVACS) or gas blow-down. In all cases, the evaporation method is slow with high risk of contamination from the solvent, glassware and blow-down gas.

Rotary evaporation

The solvent is removed under reduced pressure by mechanically rotating the flask containing the sample in a controlled temperature water bath. The (waste) solvent is condensed and collected for disposal. Problems can occur due to loss of volatile analytes, adsorption onto glassware, entrainment of analyte in the solvent vapour and the uncontrollable evaporation process.

Kuderna–Danish (K-D) evaporative concentration[3]

The Kuderna–Danish evaporative condenser was developed in the laboratories of Julius Hyman and Co., Denver, Colorado.[4] It consists of an evaporation flask (500 ml) connected at one end to a Snyder column and the other end to a concentrator tube (10 ml) (Figure 3.3). The sample containing organic solvent (200–300 ml) is placed in the apparatus, together with one or two boiling chips, and

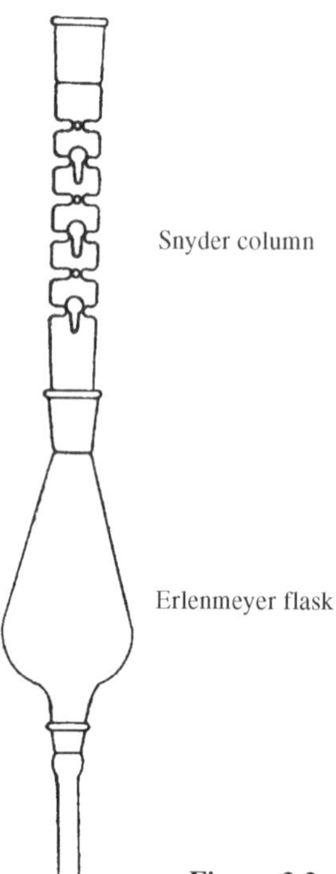

Figure 3.3 Kuderna–Danish evaporative condenser

heated with a water bath. The temperature of the water bath should be maintained at 15–20 °C above the boiling point of the organic solvent. The positioning of the apparatus should allow partial immersion of the concentrator tube in the water bath but also allow the entire lower part of the evaporation flask to be bathed with hot vapour (steam). Solvent vapours then rise and condense within the Snyder column. Each stage of the Snyder column consists of a narrow opening covered by a loose-fitting glass insert. Sufficient pressure needs to be generated by the solvent vapours to force their way through the Snyder column. Initially, a large amount of condensation of these vapours returns to the bottom of the Kuderna–Danish apparatus. In addition to continually washing the organics from the sides of the evaporation flask, the returning condensate also contacts the rising vapours and assists in the process of recondensing volatile organics. This process of solvent distillation concentrates the sample to approximately 1–3 ml in 10–20 min. Escaping solvent vapours are recovered using a condenser and collection device. The major disadvantage of this method is that violent solvent eruptions can occur in the apparatus leading to sample losses. Micro-Snyder column systems can be used to reduce the solvent volume still further.

Automated evaporative concentration system (EVACS)

Solvent from a pressure-equalised reservoir (500 ml capacity) is introduced, under controlled flow, into a concentration chamber (Figure 3.4).[5] Glass indentations regulate the boiling of solvent so that bumping does not occur. This reservoir is surrounded by a heater. The solvent reservoir inlet is situated under the level of the heater just above the final concentration chamber. The final concentration chamber is calibrated to 1.0 and 0.5 ml volumes. A distillation column is connected to the concentration chamber. Located near the top of the column are four rows of glass indentations which serve to increase the surface area. Attached to the top of the

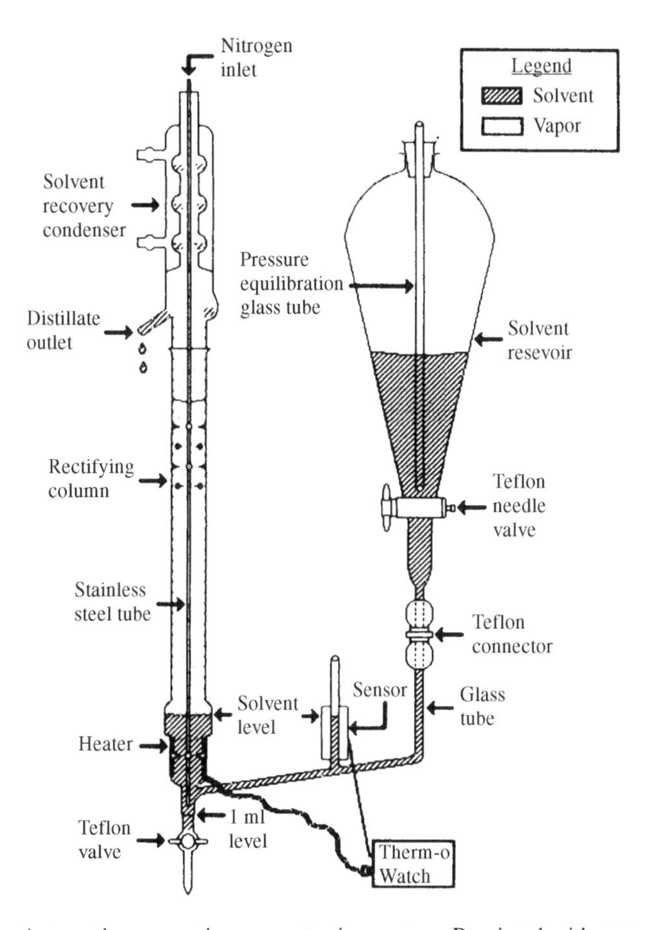

Figure 3.4 Automatic evaporative concentration system. Reprinted with permission from Ibrahim *et al. Analytical Chemistry*, **59** (1987) 2091. Copyright (1987) American Chemical Society

column is a solvent recovery condenser with an outlet to collect and hence recover the solvent.

To start a sample, the apparatus is operated with 50 ml of high purity solvent under steady uniform conditions at total reflux for 30 min to bring the system to equilibrium. Then the sample is introduced into the large reservoir either as a single volume or over several time intervals. (Note: A boiling point difference of approximately 50 °C is required between solvent and analyte for the highest recoveries.) The temperature is maintained to allow controlled evaporation. For semivolatile analytes this is typically at 5 °C higher than the boiling point of the solvent. The distillate is withdrawn while keeping the reflux ratio as high as possible. During operation, a sensor monitors the level of liquid, allowing heating to be switched off or on automatically (when liquid is present the heat is on and vice versa). After evaporation of the sample below the sensor level, the heating is switched off. After 10 min the nitrogen is started for final concentration from 10 ml to 1 ml (or less). Mild heat can be applied according to the sensitivity of solvent and analyte to thermal decomposition. When the liquid level drops below the tube, stripping nearly stops. The tube is sealed at the bottom, so that the nitrogen is dispersed above the sample and the reduction of volume becomes extremely slow. This prevents the sample from going to dryness even if left for hours. The sample is drained and the column is rinsed with two 0.5 ml aliquots of solvent. Further concentration can take place, if required.

Gas blow-down

A gentle stream of (high purity) purge gas is passed over the surface of the extract. The extract may be contained inside a conical-tipped or similar vessel. In this situation the purge gas is directed towards the side of the vessel, and not directly onto the top of the extract, to induce a swirling action. The extract-containing vessel may be partially immersed in a water bath to speed up the evaporation process. Alternative ways to speed up the evaporation process are to increase the flow rate of the impinging gas (too high a rate and losses may occur), alter its position with respect to the extract surface or increase the solvent extract surface area available for evaporation. If carry-over of trace quantities of the aqueous sample or high-boiling point solvent has occurred it may not be possible to evaporate them without significant losses of the analyte of interest. The extract may be taken to dryness or left as a small volume (~ 1 ml).

A comparison between EVACS and Kuderna–Danish (K-D) for solvent concentration during environmental analysis of trace organic chemicals has been made.[5] Selected results for a range of organic compounds (Table 3.1), lower boiling point compounds (Table 3.2) and volatile organic compounds (Table 3.3) are shown. The authors conclude that the evaporative concentration system (EVACS) is more efficient than the Kuderna–Danish evaporation for concentrating semivolatile–volatile organic analytes (bp 80–385 °C) from a low-boiling solvent (dichloromethane, bp 40 °C). The advantages of EVACS over K-D were described as follows:

- The temperature control associated with the EVACS could allow evaporation of a wider range of solvents and avoid violent boiling (bumping) which often happens with K-D.
- Fine adjustment of the temperature gives better control on compounds that are sensitive to thermal decomposition, e.g. 2,4-dichlorophenol.
- The time of operation is known for the EVACS depending on the evaporation rate needed for a certain application. This is not possible with the K-D, which boils vigorously in the beginning, then cools down during operation depending on the surrounding conditions due to lack of control of the heat supplied.
- Any volume of sample can be concentrated in the same container with continuous operation down to a 0.5–1 ml final volume.
- Sample contamination via transfer of K-D glassware is eliminated.
- The EVACS avoids the risk of losing samples by evaporation to dryness even if the system is left unattended for hours because of the spacial design of the nitrogen tube. Evaporation to dryness can happen with the K-D during the first or second step.
- The apparatus is easy to build and operate.
- Automation makes the process very easy, but even without automation, the apparatus needs only minimal attention to maintain the same level in the concentration chamber.

Table 3.1 Comparative recoveries for two-step K-D and EVACS[a] (from 200 to 1 ml)[5]

Compound	Initial concentration, (mg l^{-1})	% recovery ± SD	
		Two-step K-D	EVACS
Acetophenone	0.241	82 ± 1	93 ± 6[b]
Isophorone	0.243	85 ± 3	88 ± 4
2,4-Dichlorophenol	0.238	78 ± 7	99 ± 2[b]
Quinoline	0.201	81 ± 12	100 ± 3[b]
1-Chlorodecane	0.239	93 ± 2	91 ± 6[b]
2-Methylnaphthalene	0.174	87 ± 1	98 ± 2[b]
Biphenyl	0.226	83 ± 2	92 ± 6[b]
1-Chlorododecane	0.239	87 ± 4	96 ± 2[b]
Diacetone-L-sorbose	0.214	—	98 ± 4
Anthracene	0.159	93 ± 5	97 ± 4
Dioctylphthalate	0.211	71 ± 4	99 ± 4[b]

[a] Conditions: solvent, dichloromethane; evaporation rate, 4 ml min^{-1} for EVACS, uncontrollable for K-D; N_2 flow rate, 1 ml s^{-1}; number of determinations, 4. [b] Significant better recovery determined statistically by ANOVA at theoretical $t_{0.975}$.

Table 3.2 Comparative recoveries for lower boiling compounds[a] (from 200 to 1 ml)[5]

Compound	Initial concentration ($mg\,l^{-1}$)	% recovery ± SD	
		K-D	EVACS
Ethyl butyrate	0.27	70 ± 2	70 ± 3
Ethylbenzene	0.29	72 ± 2	71 ± 3
Cyclohexanone	0.29	74 ± 1	76 ± 4
Anisole	0.29	72 ± 1	76 ± 3[b]
1,4-Dichlorobenzene	0.27	75 ± 2	80 ± 2[b]
2-Ethylhexanol	0.26	76 ± 4	82 ± 3[b]
Tolunitrile	0.27	82 ± 3	80 ± 1
Naphthalene	0.26	88 ± 2	96 ± 3[b]
Benzothiazole	0.27	100 ± 1	99 ± 3
Ethyl cinnamate	0.26	90 ± 1	98 ± 6

[a] Conditions: solvent, dichloromethane; evaporation rate, $4\,ml\,min^{-1}$ for EVACS, uncontrollable for K-D; N_2 flow rate, $1\,ml\,s^{-1}$; number of determinations, 4. [b] Significant better recovery determined statistically by ANOVA at theoretical $t_{0.975}$.

Table 3.3 Comparative recoveries for volatile organic compounds[a] (from 200 to 1 ml)[5]

Compound	Boiling point (°C)	Initial concentration ($mg\,l^{-1}$)	% Recovery ± SD	
			K-D	EVACS[b]
Trichloroethylene	87.0	0.40	17 ± 3	24 ± 1
Benzene	80.1	0.52	29 ± 2	38 ± 1
Perchloroethylene	121.14	0.40	53 ± 3	63 ± 4
Toluene	110.6	0.29	53 ± 1	68 ± 6
Chlorobenzene	132.22	0.41	55 ± 3	68 ± 1

[a] Conditions: solvent, dichloromethane; evaporation rate, $2\,ml\,min^{-1}$ for EVACS, uncontrollable for K-D; N_2 flow rate, $1\,ml\,s^{-1}$; number of determinations, 3. [b] Significant better recovery determined statistically by ANOVA at theoretical $t_{0.975}$ for all compounds.

- Economically, EVACS has the advantage that solvents are recovered for reuse for similar samples especially when evaporating large volumes of solvents without disturbing the process.

In addition, the EVACS procedure has been designed to avoid most of the difficulties and weaknesses associated with the conventional K-D method. Possible errors that are still associated with the EVACS and K-D methods can result from the following:

- Cross contamination can occur between samples. Careful cleaning is required between samples.
- Possibility of thermal decomposition of certain compounds, which therefore require fine adjustment of heat especially during nitrogen stripping.
- Quantitative error can result from the manual addition of internal standards into the final volume prior to analysis.
- Instrumental analytical errors. These can be minimised by the addition of internal standards.

3.2 PURGE AND TRAP FOR VOLATILE ORGANICS

Purge and trap is widely used for the extraction of volatile organic compounds from aqueous samples followed by gas chromatography. The method involves the introduction of an aqueous sample (typically 5 ml) into a glass sparging vessel (Figure 3.5). The sample is then purged with (high purity) nitrogen at a specified flow rate and time. The extracted volatile organics are then transferred to a trap, e.g.

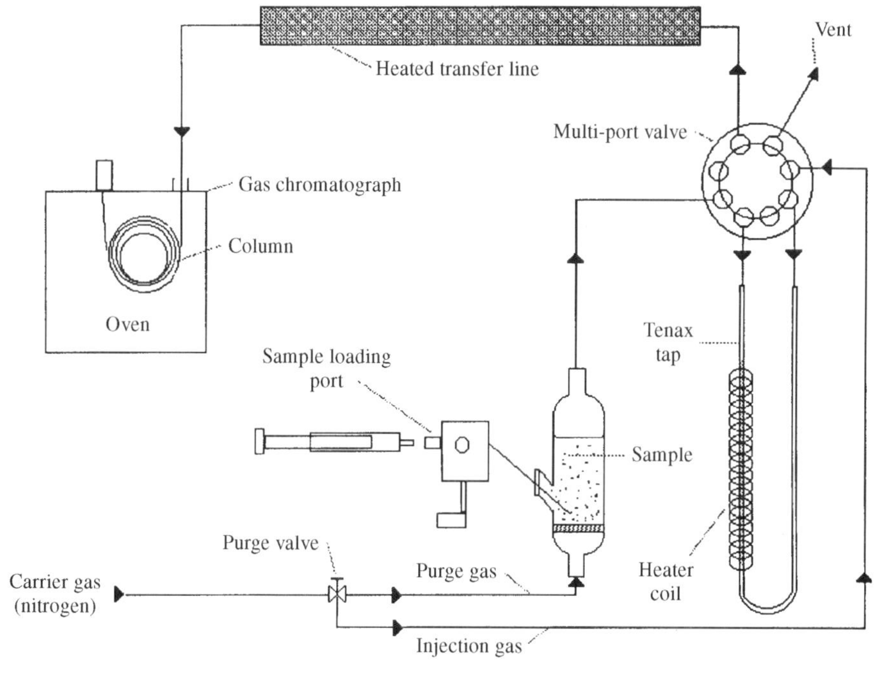

Figure 3.5 Purge and trap system for gas chromatography. Reproduced by permission of Mr E. Ludkin, University of Northumbria at Newcastle

Tenax, at ambient temperature. This is followed by the desorption step. In this step, the trap is rapidly heated to desorb the trapped volatile organic compounds in a narrow band. The desorbed compounds are transferred via a heated transfer line to the injector of a gas chromatograph for separation and detection. A recent review has highlighted the development of the purge and trap technique.[6]

REFERENCES

1. M.S. Cresser, *Solvent Extraction in Flame Spectroscopic Analysis*, Butterworths, London (1978).
2. R.E. Majors, *LC-GC Int.*, **10**(2) (1997) 93.
3. F.W. Karasek, R.E. Clement and J.A. Sweetman, *Anal. Chem.*, **53** (1981) 1050A.
4. F.A. Gunther, R.C. Blinn, M.J. Kolbezen and J.H. Barkley, *Anal. Chem.*, **23** (1951) 1835.
5. E.A. Ibrahim, I.H. Suffet and A.B. Sakla, *Anal. Chem.*, **59** (1987) 2091.
6. S.M. Abel, A.K. Vickers and D. Decker, *J. Chromatogr. Sci.*, **32** (1994) 328.

4

Solid Phase Extraction

Solid phase extraction (SPE) or sometimes referred to as liquid–solid extraction, involves bringing a liquid or gaseous sample in contact with a solid phase or sorbent whereby the analyte is selectively adsorbed onto the surface of the solid phase.[1] The solid phase is then separated from the solution and other solvents (liquids or gases) added. The first such solvent is usually a wash to remove possible adsorbed matrix components; eventually an eluting solvent is brought into contact with the sorbent to selectively desorb the analyte. The focus of this chapter will be on SPE with liquid samples and solvents. The solid phase sorbent is usually packed into small tubes or cartridges and resembles a small liquid chromatography column. Recently the sorbent has become available in round, flat sheets that can be mounted in a filtration apparatus. In this case the sorbent resembles that of the commonly used filter paper. Whichever design is used the sample-containing solvent is forced by pressure or vacuum through the sorbent. By careful selection of the sorbent, the analyte should be retained by the sorbent in preference to other extraneous material present in the sample. This extraneous material can be washed from the sorbent by the passing of an appropriate solvent. Subsequently the analyte of interest can then be eluted from the sorbent using a suitable solvent. This solvent is then collected for analysis. Obviously further sample clean-up or preconcentration can be carried out, if desired.

From this brief introduction the reader should realise that the choice of sorbent and the solvent system used is of paramount importance for effective preconcentration and/or clean-up of the analyte in the sample. This process of SPE should allow more effective detection and identification of the analyte.

4.1 TYPES OF SPE MEDIA

Generally SPE sorbents can be divided into three classes; normal phase, reversed phase and ion exchange. The most common sorbents are based on silica particles (irregular shaped particles with a particle diameter between 30 and 60 μm) to which functional groups are bonded to surface silanol groups to alter their retentive properties (it should be noted that unmodified silica is sometimes used). The bonding of the functional groups is not always complete, so unreacted silanol groups remain. These unreacted sites are polar, acidic sites and can make the interaction with analytes more complex. To reduce the occurence of these polar sites, some SPE media are 'end-capped', that is a further reaction is carried out on the residual silanols using a short-chain alkyl group. End-capping is not totally effective. It is the nature of the functional groups that determine the classification of the sorbent. In addition to silica some other common sorbents are based on florisil, alumina and macroreticular polymers.

Normal phase sorbents have polar functional groups, e.g. cyano, amino and diol (also included in this category is unmodified silica). The polar nature of these sorbents means that it is more likely that polar compounds, e.g. phenol, will be retained. In contrast, reversed phase sorbents have non-polar functional groups, e.g. octadecyl, octyl and methyl, and conversely are more likely to retain non-polar compounds, e.g. polycyclic aromatic hydrocarbons. Ion exchange sorbents have either cationic or anionic functional groups and when in the ionized form attract compounds of the opposite charge. A cation exchange phase, such as benzenesulfonic acid, will extract an analyte with a positive charge (e.g. phenoxyacid herbicides) and vice versa. A summary of commercially available silica-bonded sorbents is given in Table 4.1.

4.2 CARTRIDGE OR DISK FORMAT

The design of the SPE device can vary, with each design having its own advantages related to the number of samples to be processed and the nature of the sample and its volume. The most common arrangement is the syringe barrel or cartridge. The cartridge itself is usually made of polypropylene (although glass and polytetrafluorethylene, PTFE, are also available), with a wide entrance, through which the sample is introduced, and a narrow exit (male luer tip). The appropriate sorbent material, ranging in mass from 50 mg to 10 g, is positioned between two frits, at the base (exit) of the cartridge, which act to both retain the sorbent material and to filter out particulate matter. Typically the frit is made from polyethylene with a 20 μm pore size.

Solvent flow through a single cartridge is typically done using a side-arm flask apparatus (Figure 4.1), whereas multiple cartridges can be simultaneously processed (from 8 to 30 cartridges) using a commercially available vacuum manifold (Figure 4.2). A variation on this type of cartridge system or syringe filter is when a plunger is inserted into the cartridge barrel. In this situation the solvent is added to the syringe barrel and forced through the SPE using the plunger. This

Table 4.1 Some commonly available silica-bonded sorbents (adapted from Reference 1)

	Phase	Bonded moiety
Nonpolar phases	C1, methyl	$Si\text{-}CH_3$
	C8, octyl	$Si\text{-}(CH_2)_7\text{-}CH_3$
	C18, octadecyl	$Si\text{-}(CH_2)_{17}\text{-}CH_3$
Polar phases	Si, silica	$Si\text{-}OH$
	CN, cyanopropyl	$Si\text{-}CH_2\text{-}CH_2\text{-}CH_2\text{-}CN$
	2OH, diol	$Si\text{-}CH_2\text{-}CH_2\text{-}CH_2\text{-}O\text{-}CH_2\text{-}CHOH\text{-}CH_2OH$
Ion-exchange phases	SCX, benzenesulfonic acid	$Si\text{-}CH_2\text{-}CH_2\text{-}CH_2\text{-}C_6H_4\text{-}SO_3^-$
	DEA, diethylammoniopropyl tertiary amine	$Si\text{-}CH_2\text{-}CH_2\text{-}CH_2\text{-}NH^+\text{-}(CH_2\text{-}CH_3)_2$
	SAX, trimethylammoniopropyl quaternary amine	$Si\text{-}CH_2\text{-}CH_2\text{-}CH_2\text{-}N^+\text{-}(CH_3)_3$

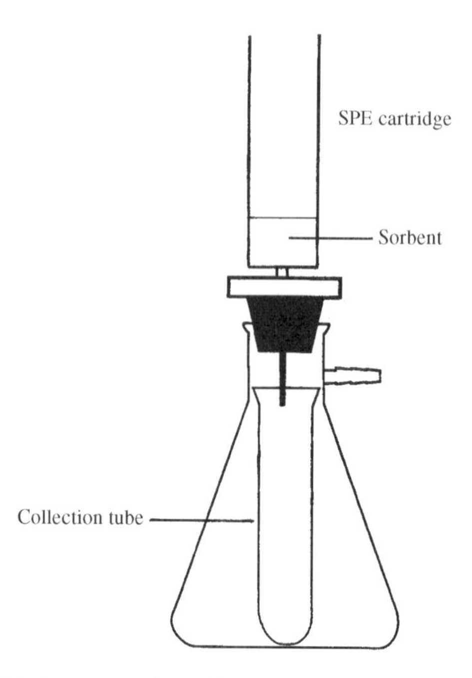

Figure 4.1 Solid phase extraction using a cartridge and a single side-arm flask apparatus

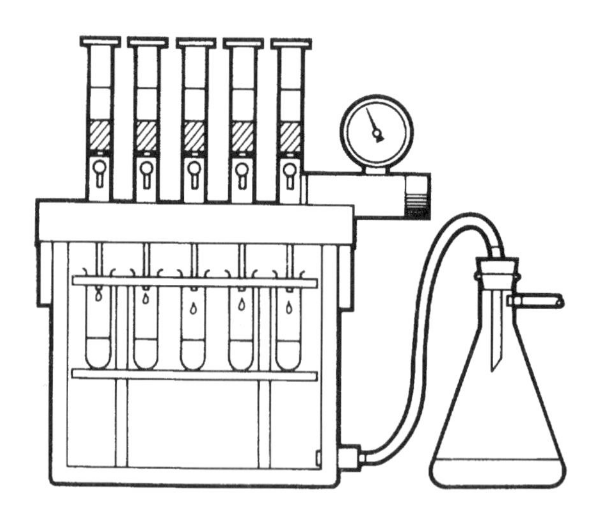

Figure 4.2 Vacuum manifold for solid phase extraction of multiple cartridges. Reproduced by permission of International Sorbent Technology Ltd

system is effective if only a few samples are to be processed, for early method development, the SPE method is simple or no vacuum system is available.

The most distinctly different approach to SPE is the use of a disk, not unlike a common filter paper. This SPE disk format is referred to by its trade name of Empore disks. The 5–10 μm sorbent particles are intertwined with fine threads of PTFE which results in a disk approximately 0.5 mm thick and a diameter in the range 47 to 70 mm. Empore disks are placed in a typical solvent filtration system and a vacuum applied to force the solvent-containing sample through (Figure 4.3). To minimise dilution effects that can occur it is necessary to introduce a test-tube into the filter flask to collect the final extract. Manifolds are commercially available for multiple sample extraction using Empore disks.

Both the cartridge and disk formats have their inherent advantages and limitations. For example the SPE disk, with its thin sorbent bed and large surface area, allows rapid flow rates of solvent. Typically, one litre of water can be passed through an Empore disk in approximately 10 min whereas with a cartridge system the same volume of water may take approximately 100 min! However, large flow rates can result in poor recovery of the analyte of interest due to there being a shorter time for analyte-sorbent interaction.

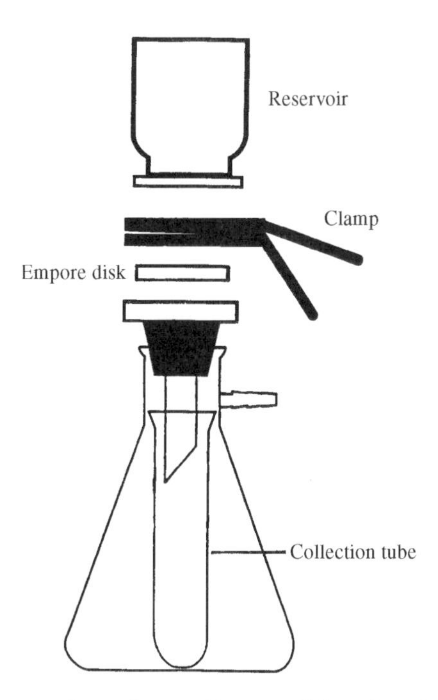

Figure 4.3 Solid phase extraction using an Empore disk and a single side-arm flask apparatus

4.3 METHOD OF SPE OPERATION

Irrespective of SPE format the method of operation is the same and can be divided into five steps (Figure 4.4).[1] Each step is characterised by the nature and type of solvent used which in turn is dependent upon the characteristics of the sorbent and the sample. The five steps are as follows: wetting the sorbent, conditioning of the sorbent, loading of the sample, rinsing or washing the sorbent to elute extraneous material, and finally elution of the analyte of interest. Wetting the sorbent allows the bonded alkyl chains, which are twisted and collapsed on the surface of the silica, to be solvated so that they spread open to form a bristle. This ensures good contact between the analyte and the sorbent in the adsorption of the analyte step. It is also important that the sorbent remains wet in the following two steps or poor recoveries can result. This is followed by conditioning of the sorbent in which solvent or buffer, similar to the test solution that is to be extracted, is pulled through the sorbent. This is followed by sample loading where the sample is forced through the sorbent material by suction, a vacuum manifold or a plunger. By careful choice of the sorbent, it is anticipated that the analyte of interest will be retained by the sorbent in preference to extraneous material and other related compounds of interest that may be present in the sample. Obviously this ideal situation does not always occur and compounds with similar structures will undoubtedly be retained also. This process is followed by washing with a suitable solvent that allows unwanted extraneous material to be removed without influencing the elution of the analyte of interest. This step is obviously the key to the whole process and is dependent upon the analyte of interest and its interaction with the sorbent material and the choice of solvent to be used. Finally the analyte of interest is eluted from the sorbent using the minimum amount of solvent to affect quantitative release. By careful control of the amount of solvent used in the elution step and the sample volume initially introduced onto the sorbent a preconcentration of the analyte of interest can be affected. Successful SPE obviously requires careful consideration of the nature of the SPE sorbent, the solvent systems to be used and their influence on the analyte of interest. In addition, it may be that it is not a single analyte that you are seeking to preconcentrate but a range of analytes. If they have similar chemical structures then a method can be successfully developed to extract multiple analytes. While this method development may seem to be laborious and extremely time-consuming it should be remembered that multiple vacuum manifolds are commercially available as are robotic systems that can carry out the entire SPE process. Once developed the SPE method can then be used to process large quantities of sample with good precision.

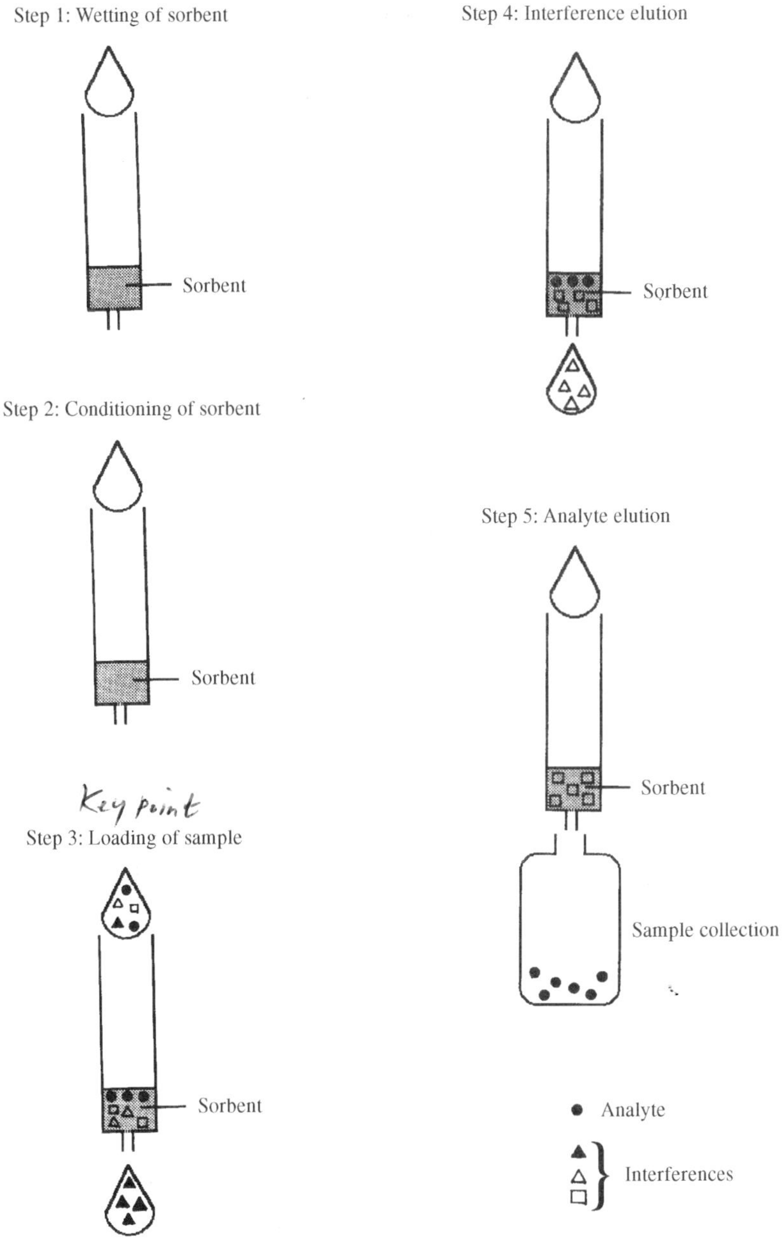

Step 1: Wetting of sorbent

Sorbent

Step 2: Conditioning of sorbent

Sorbent

Step 3: Loading of sample

Sorbent

Step 4: Interference elution

Sorbent

Step 5: Analyte elution

Sorbent

Sample collection

● Analyte

▲ △ □ } Interferences

Figure 4.4 Solid phase extraction: method of operation

4.4 SOLVENT SELECTION

The choice of solvent directly influences the retention of the analyte on the sorbent and its subsequent elution, whereas the solvent polarity determines the solvent strength (or ability to elute the analyte from the sorbent in a smaller volume than a weaker solvent). The solvent strength for normal- and reversed-phase sorbents is shown in Table 4.2. Obviously this is the ideal. In some situations it may be that no individual solvent will perform its function adequately so it is possible to resort to a mixed solvent system. It should also be noted that for a normal-phase solvent, both solvent polarity and solvent strength are coincident whereas this is not the case for a reversed-phase sorbent. In practice, however, the solvents normally used for reversed-phase sorbents are restricted to water, methanol, isopropyl alcohol and acetonitrile. For ion-exchange sorbents, solvent strength is not the main effect; pH and ionic strength are the main factors governing analyte retention on the sorbent and its subsequent elution. As with the choice of sorbent some preliminary work is required to affect the best solvents to be used. Using a reversed-phase sorbent (e.g. C18) as an example the general methodology to be followed for SPE is described.

Sorbent: C18
Wetting the sorbent: Pass 1.0 ml of methanol or acetonitrile per 100 mg of sorbent. This solvent has several functions, it will remove impurities from the sorbent that may have been introduced in the manufacturing process. In addition, as reversed-phase sorbents are hydrophobic, they need the organic solvent to solvate or wet their surfaces.
Conditioning: Pass 1 ml of water or buffer per 100 mg of sorbent. Do not allow the sorbent to dry out before applying the sample.

Table 4.2 Solvent strengths for normal- and reversed-phase sorbents

Solvent strength for normal phase sorbents		Solvent strength for reversed phase sorbents
Weakest	Hexane	**Strongest**
	Iso-octane	
	Toluene	
	Chloroform	
	Dichloromethane	
	Tetrahydrofuran	
	Ethyl ether	
	Ethyl acetate	
	Acetone	
	Acetonitrile	
	Isopropyl alcohol	
Strongest	Methanol	
	Water	**Weakest**

Loading: A known volume of sample is loaded in a high polarity solvent or buffer. The solvent may be one that has been used to extract the analyte from a solid matrix.
Rinsing: Unwanted, extraneous material is removed by washing the sample-containing sorbent with a high-polarity solvent or buffer. This process may be repeated.
Elution: Elute analytes of interest with a less polar solvent, e.g. methanol or the HPLC mobile phase (if this is the method of subsequent analysis). 0.5–1.0 ml per 100 mg of sorbent is typically required for elution.

Finally, the SPE cartridge or disk is discarded.

4.5 FACTORS AFFECTING SPE

While the choice of SPE sorbent is highly dependent upon the analyte of interest and the sorbent system to be used, certain other parameters can influence the effectiveness of the SPE methodology. Obviously the number of active sites available on the sorbent cannot be exceeded by the number of molecules of analyte otherwise breakthrough will occur. Therefore, it is important to assess the capacity of the SPE cartridge or disk for its intended application. In addition, the flow rate of sample through the sorbent is important; too fast a flow and this will allow minimal time for analyte-sorbent interaction. This must be carefully balanced against the need to pass the entire sample through the cartridge or disk. It is normal therefore for an SPE cartridge to operate with a flow rate of 3–10 ml min^{-1} whereas 10–100 ml min^{-1} are typical for the disk format.

Once the analyte of interest has been adsorbed by the sorbent, it may be necessary to wash the sorbent of extraneous matrix components prior to elution of the analyte. Obviously the choice of solvent is critical in this step, as has been discussed previously. For the elution step it is important to consider the volume of solvent to be used (as well as its nature). For quantitative analysis, by, for example HPLC or GC, two factors are important: (i) preconcentration of the analyte of interest from a relatively large volume of sample to a small extract volume and, (ii) clean-up of the sample matrix to produce a particle-free and chromatographically clean extract. All of these factors require some method development either using a trial-and-error approach or by consultation with existing literature. It is probable that both are required in practice.

4.6 SELECTED METHODS OF ANALYSIS FOR SPE

The general methodology to be followed for off-line SPE will be described using selected literature examples with emphasis on a reversed-phase system (C18).

Analyte: selected organochlorine pesticides
Sorbent: C18 cartridge
Wetting/conditioning the sorbent: 2×5 ml of methanol followed by 2×2 ml of water
Loading: 1 litre of drinking and surface water samples from the Gdansk district, Poland were passed at a flow rate of approximately 10 ml min^{-1}
Rinsing: none
Elution: 2×2.5 ml of n-pentane and dichloromethane (1 : 1, v/v)
Other comments: both extracts were combined and evaporated to 0.5 ml prior to GC-ECD analysis
Reference: Biziuk *et al.*[2]

Analyte: organochlorine pesticides
Sorbent: C18 or C8, Empore disk
Wetting/conditioning the sorbent: 10 ml of methanol followed by 10 ml of distilled water.
Loading: 0.5 litre of spiked water sample containing 2.5% methanol
Rinsing: none
Elution: 5 ml of ethyl acetate and 5 ml of hexane
Other comments: extract was evaporated to 0.2 ml under a stream of nitrogen at 45 °C prior to GC-ECD analysis.
Reference: Viana *et al.*[3]

Analyte: 26 organophosphorus, organochlorine and other pesticides
Sorbent: C18 cartridge
Wetting/conditioning the sorbent: consecutive additions of 5 ml of isooctane, 5 ml of ethyl acetate, 5 ml of methanol and 10 ml of deionized water
Loading: 1 litre of spiked water sample, pH 6.5, added at a flow rate of 10–15 ml min^{-1} under vacuum
Rinsing: 10 ml of deionized water and dried by aspirating air for 30 min
Elution: 3×0.5 ml of ethyl acetate and 3×0.5 ml of isooctane
Other comments: extract was dried over anhydrous sodium sulfate and washed with an additional 0.5 ml of each eluting solvent. The combined extracts were concentrated to dryness under a gentle stream of nitrogen. The final residue was dissolved in 1 ml of hexane and internal standard added prior to GC-ECD or FPD analysis
Reference: de la Colina *et al.*[4]

Analyte: fenamiphos and metabolites
Sorbent: C18 cartridge
Wetting/conditioning the sorbent: 5 ml of ethyl acetate and 5 ml of methanol were added without vacuum
Loading: 100 ml of spiked mineral water at pH 7 added under vacuum
Rinsing: none

Elution: sorbent dried for 15–20 min and then eluted with 5 ml of ethyl acetate
Other comments: extract evaporated under a stream of nitrogen prior to GC-MSD analysis
Reference: Terreni *et al.*[5]

Analyte: ten carbamate pesticides
Sorbent: C18, Empore disks
Wetting/conditioning the sorbent: 10 ml of methanol under vacuum, then 10 ml of acetonitrile (after drying). Subsequently 30 ml of water ensuring that the disk does not become dry prior to addition of sample
Loading: 2 litres of Erbo river water
Rinsing: none
Elution: After drying for 1 hour under vacuum, elution was performed using 2×10 ml of acetonitrile
Other comments: extract was evaporated to dryness and the residue diluted in 500 μl of methanol prior to analysis by HPLC-MS.
Reference: Honing *et al.*[6]

Analyte: alachlor
Sorbent: C18, Empore disks
Wetting/conditioning the sorbent: 10 ml of ethyl acetate was passed and then disk dried for 2 min. Then, 10 ml of acetone was added and the disks were dried for 5 min. Afterwards, 15 ml of methanol was added and when film of methanol was very thin, 10×10 ml of deionized water was added. Solvents were drawn through the disk at a rate of 0.5 ml s^{-1}
Loading: 25–300 ml of spiked water sample. Sample was drawn through the disk at a rate of 0.5 ml s^{-1}. After sample application the vacuum was left on for 30 min to allow the disk to dry
Rinsing: none
Elution: 3×20 ml portions of ethyl acetate
Other comments: extract reduced in volume, using a rotary evaporator, to 0.1–1.0 ml prior to analysis by GC-ECD and/or GC-MSD
Reference: Penuela and Barcelo[7]

Analyte: eight polar acid herbicides
Sorbent: C18 cartridge
Wetting/conditioning the sorbent: 10 ml of methanol followed by 10 ml of deionized water acidified to pH 2.4–2.6 with HCl. The cartridge was not allowed to run dry during this process
Loading: 500 ml of surface water sample acidified to pH 2.4–2.6 with 2 M HCl and containing 5 ml of methanol and 1.0 ml of internal standard was percolated through the cartridge under vacuum at a rate of approximately 15 ml min^{-1}. Cartridge was then air-dried under vacuum for 5 min and centrifuged for 15 min at 500 g to remove the remaining water

Rinsing: The cartridge was then washed with 0.75 ml of methanol and the solution discarded
Elution: 2×2 ml of methanol, under vacuum, at a rate of approximately 2 ml min^{-1}
Other comments: extract was then adjusted to 10 ml with methanol followed by derivatisation and column chromatographic clean-up prior to GC analysis
Reference: Vink and van der Poll[8]

Analyte: polar thermally labile pesticides
Sorbent: C18
Wetting/conditioning the sorbent: 5 ml of acetone, 5 ml methanol and 5 ml of deionized water
Loading: 1 litre of sample extracted at a flow of 9–10 ml min^{-1} (using a manifold system that allows up to 12 cartridges to be extracted in parallel). Cartridge was then dried for 30 min using a gentle stream of nitrogen
Elution: 5×1 ml of methanol (more effective than 1×5 ml of methanol)
Other comments: extract reduced in volume, under a stream of nitrogen, to 1 ml for GC and 0.5 ml for HPLC. Internal standard added to this final sample volume. 1 ml of water added to extract for HPLC analysis
Reference: Eisert *et al.*[9]

4.7 AUTOMATED AND ON-LINE SPE

The use of automated SPE allows large numbers of samples to be extracted routinely, with unattended operation of a system. The use of automated SPE should therefore allow more samples to be extracted (higher sample throughput) with better precision. In addition, it also allows the analyst to perform other tasks or prepare more samples for analysis. Two categories of automated SPE can be distinguished: the use of instrumentation that imitates the manual off-line procedure and an on-line SPE procedure that utilises column switching. The former approach imitates the off-line manipulations required for SPE via a robotic arm or autosampler. Thus it is possible to programme the steps of SPE, i.e. wetting, conditioning, sample loading, washing and elution, and then collect the analyte in an appropriate solvent. The volumes to be used for each step are programmed into the system as a method. This assumes that the SPE method has been previously well characterised. After completion of this process the extracted analyte is ready for chromatographic analysis.

On-line SPE is the situation where the eluent of the SPE column is automatically directed into the chromatograph (assuming it to be HPLC, although this is not always the case) for separation and quantitation of the analytes of interest. This situation is often described as 'column switching' or 'coupled column' techniques. The SPE column or 'pre-column' frequently contains a low efficiency sorbent which performs a preseparation of the sample, after which the analyte-containing fraction is directed onto a second high efficiency column for separation and quantitation of the analytes of interest. A simplified diagram for column switching is shown in Figure 4.5. The solvent to wet and precondition the sorbent is pumped through the pre-column and then directed to waste. Then the sample is loaded onto the pre-column and rinsed with an appropriate solvent. In the elution step, the high pressure switching valve is rotated so that the mobile phase passes through the pre-column and flushes the analytes onto the analytical separation column. While the analytical separation takes place the switching valve returns to the 'load' position for reconditioning of the pre-column ready to start the next sample. Commercial systems are available that utilise this automated on-line procedure, e.g. the Prospekt system from Spark Holland.

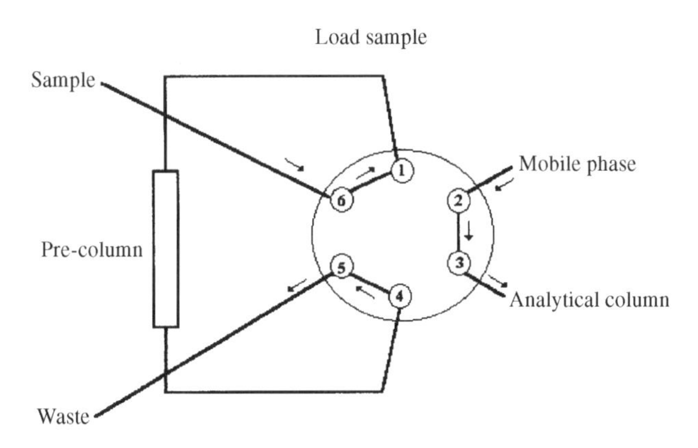

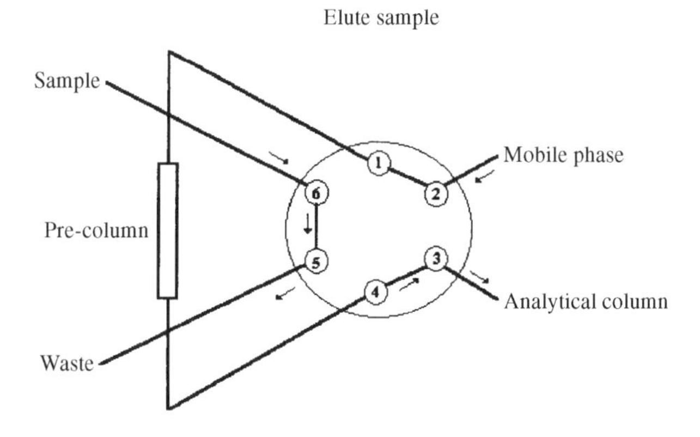

Figure 4.5 Schematic diagram illustrating the principle of column switching

The use of on-line SPE offers several advantages to the laboratory: the number of manual manipulations decreases which improves the precision of the data, there is a lower risk of contamination as the system is closed from the point of sample injection through to the chromatographic output to waste, all of the analyte loaded onto the pre-column is transferred to the analytical column, and the analyst is available to perform other tasks. These advantages must, of course, be balanced by some disadvantages: the initial time taken to develop a method that is both robust and reliable in terms of both the column technology (pre-column and analytical column) and the equipment used, and the additional capital cost involved. These are important considerations. It is envisaged that off-line SPE is the preferred method of choice for non-routine samples whereas an automated on-line SPE system would be used for large numbers of routine samples, process monitoring and the monitoring of dynamic systems. Selected applications of automated SPE will now be used to highlight the diversity of applications to which this technology has already been applied.

4.8 SELECTED APPLICATIONS OF AUTOMATED ON-LINE SPE

The continuous monitoring of low levels of pollutants in the river Rhine has been a focus for the development of on-line monitoring systems.[10] In 1990, a major research initiative commenced, aimed at developing methods to continuously monitor pollutants in the river Rhine and other European rivers. This so-called Rhine Basin Programme is managed by Hewlett-Packard's analytical division in Waldbronn, Germany and involves seven other participants:

- The Engler Bunte Institute at the University of Karlsruhe, Germany in conjunction with the research arm of the Deutsche Vereins des Gas und Wasserfaches, an association of gas and water specialists. Both groups focus on water analysis and treatment.
- The Environmental Office of the German state of North Rhine Westphalia, in Dusseldorf, which is in charge of monitoring water quality and assessing the health risks of chemicals in the water and its segment of the Rhine.
- The Free University of Amsterdam's Department of Analytical Chemistry, which focuses much of its research on physical separation methods, including three facets of chromatography: column-liquid; capillary gas, and supercritical fluid techniques.
- The Netherlands Institute for Inland Water Management and Wastewater Treatment, in Lelystad. This national agency, which operates its own laboratories, gives technical and scientific support to officials responsible for surface waters.
- The Institute for Water Supply, Wastewater Cleanup and Flood Protection, in Dubendorf, Switzerland. This Swiss federal institute focuses on interdisciplinary research projects on water quality, including chemical analysis of pollutants.

- The Societé Anonyme de Gestion des Eaux de Paris–the communal drinking-water company of Paris, which is interested in continuously monitoring water for potentially harmful compounds.
- Acer Environmental, a UK environmental consultancy that is part of Welsh Water. Acer has developed new methods for screening water quality and continuous monitoring of rivers.

Several compound types were chosen as priority areas: polar pesticides (agricultural usage), phthalate esters (widely used as plasticisers in PVC), fluorescent whitening agents (used mainly in textile and paper manufacturing), organophosphorus compounds (crop protection agents), organic sulfonic acids (used widely in the dye and detergent industries) and organotin compounds (used as agricultural fungicides and as anti-fouling agents in ships' paint). Some of these analytical methods have been incorporated into fully automated monitoring systems called SAMOS (**S**ystem for the **A**utomatic **M**easurement of **O**rganic micropollutants in **S**urface water).

An automated SAMOS-HPLC system has been developed based on on-line SPE. In this system, shown in Figure 4.6, a disposable SPE pre-column (typically 5–10 mm length × 2–3 mm i.d.) containing either a C18 bonded silica or a more hydrophobic styrene-divinylbenzene copolymer, such as PLRP-S is used to trap the analytes from 100–ml of the aqueous sample, loaded at a rate of 5–10 ml min^{-1}. After a rapid clean-up with a small volume of high purity water, desorption is performed by coupling the pre-column on-line with the analytical column. Separation is achieved using an alkyl-bonded silica column, and a gradient mobile phase of acetonitrile-aqueous phosphate buffer. Detection is achieved using a photodiode array monitoring at wavelengths of 210/220/245/280 nm or 220/230/245/270/300 nm. An example chromatogram showing the detection of 27 polar pesticides, spiked at three concentration levels, in Amsterdam tap water is shown in Figure 4.7. As the European Union alert and alarm levels for individual pesticides in surface waters are set at 1 $\mu g\,l^{-1}$ and 3 $\mu g\,l^{-1}$, respectively it is seen that adequate sensitivity is achievable in spite of the large background hump, that occurs due to the presence of humic and fulvic acids, near the start of the chromatogram. Various examples of the use of this system have been reported and selected chromatograms demonstrate the applicability of the system.

Example 1 Glyphosate (Gly) and its main metabolite, aminomethylphosphonic acid (Ampa) in water using on-line HPLC with fluorescence detection[11]

Pretreatment: A 300 μl sample is buffered to pH 8.7 with borate buffer. Then the derivatising agent, FMOC added. After mixing and a reaction time of 20 min, phosphoric acid is added to stop the reaction. After mixing, an aliquot of the mixture is injected for on-line SPE.

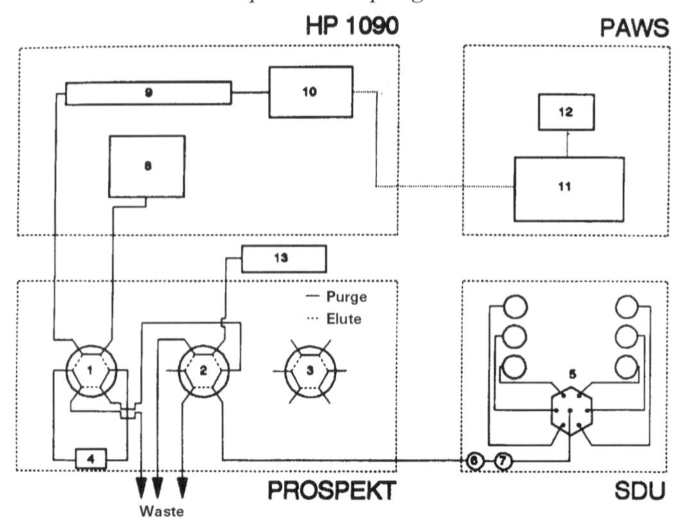

Figure 4.6 Schematic diagram of an automated SAMOS-HPLC system. (1,2,3 = high pressure valves of the Prospekt; 4 = trace-enrichment cartridge of Prospekt; 5 = solenoid value; 6 = pulse damper; 7 = purge valve; 8 = solvent delivery system of HP 1090 liquid chromatograph; 9 = analytical column; 10 = diode array detector; 11 = Pascal workstation (PAWS); 12 = printer; 13 = preparative pump for sample loading; SDU = solvent delivery unit.) Reprinted from Brouwer *et al. Journal of Chromatography*, **703** (1995) 167, with kind permission of Elsevier Science-NL, Sara Burgerhartstraat 25, 1055 KV Amsterdam, The Netherlands

SPE: Sorbent used was a proprietary material, C-40. After activation and conditioning of the sorbent, the derivatised sample is loaded using 2-propanol for both the transfer and the clean-up. After clean-up the SPE is switched on-line with the HPLC.
Analytical separation and quantitation: Aminopropyl column with an isocratic mobile phase (75/25 v/v%, acetonitrile–0.1 v/v% phosphoric acid) at a flow rate of $1\,ml\,min^{-1}$. Fluorescence detection was done at an excitation wavelength of 265 nm and an emission wavelength of 300 nm.

An example chromatogram is shown in Figure 4.8 of a blank and water spiked at $2\,\mu g\,l^{-1}$.

Example 2 Pesticides in drinking water[12]

SPE: PRLP-S, 10×2.0 mm i.d.
Analytical separation and detection: ODS, 25×0.46 cm i.d., gradient mobile phase (acetonitile–aqueous phosphate buffer (pH 7)) at a flow rate of $1\,ml\,min^{-1}$. Ultraviolet detection at 220, 249 and 268 nm.

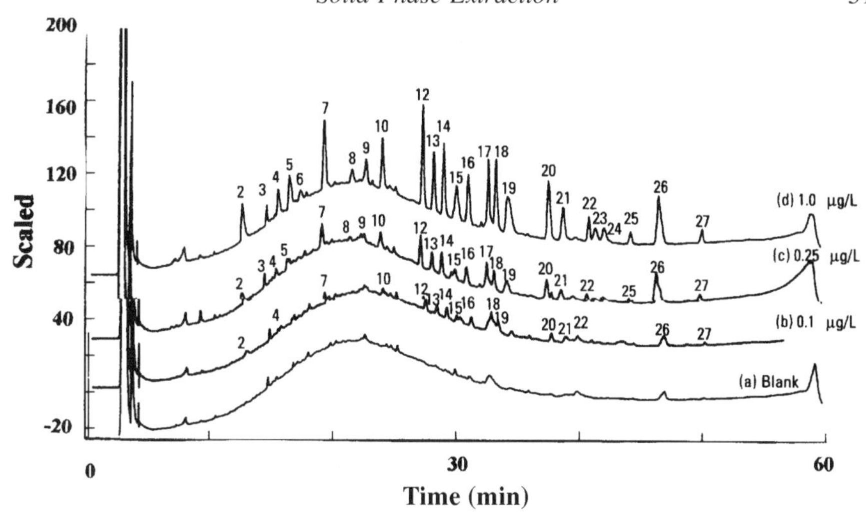

Figure 4.7 Detection of 27 polar pesticides using SPE-HPLC diode array detection. Analysis of 100 ml of Amsterdam tap water. (a) Raw tap water spiked at levels of (b) $0.1\,\mu g\,l^{-1}$; (c) $0.25\,\mu g\,l^{-1}$; and (d) $1.0\,\mu g\,l^{-1}$. Peak assignment: 1 = aniline; 2 = carbendazim; 3 = 1-(3-chloro-4-hydroxyphenol)-3,3-dimethylurea; 4 = metamitron; 5 = chloridiazone; 6 = dimethoate; 7 = monomethyl metoxuron; 8 = aldicarb; 9 = bromacil; 10 = cyanazine; 11 = 2-nitrophenol; 12 = chlorotoluron; 13 = atrazine; 14 = diuron; 15 = metobromuron; 16 = metazachlor; 17 = propazine; 18 = warfarin; 19 = 3,3′-dichlorobenzidine; 20 = barban; 21 = alachlor; 22 = nitralin; 23 = dinoseb; 24 = dinoterb; 25 = phoxim; 26 = nitrofen; and 27 = trifluralin. Reprinted with permission from Brinkman *Environmental Science and Technology*, **29** (1994) 80A. Copyright (1995) American Chemical Society

Example chromatogram (Figure 4.9) of 150 ml of drinking water sample spiked with $0.3\,\mu g\,l^{-1}$ pesticides.

Example 3 Triazines in drinking water[12]

SPE: PRLP-S, 10 × 2.0 mm i.d.
Analytical separation and detection: ODS, 25 × 0.46 cm i.d., gradient mobile phase (acetonitile-aqueous phosphate buffer (pH 7)) at a flow rate of $1\,ml\,min^{-1}$. Ultraviolet detection at 220 nm.

Example chromatogram (Figure 4.10) of 300 ml of drinking water sample spiked with $0.1\,\mu g\,l^{-1}$ triazines.

Example 4 Phenyl urea herbicides in drinking water[12]

SPE: PRLP-S, 10 × 2.0 mm i.d.

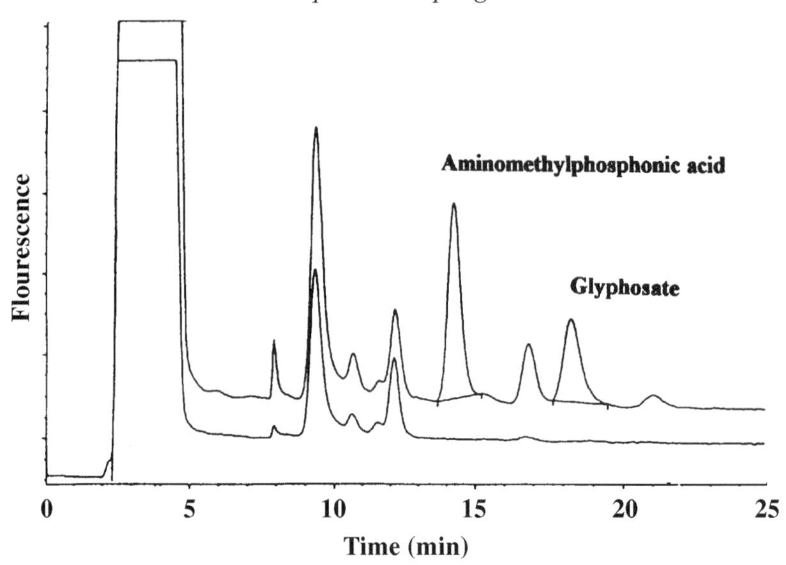

Figure 4.8 Chromatogram of blank and water spiked with glyphosate and aminomethylphosphonic acid (metabolite of glyphosate) at 2 $\mu g\,l^{-1}$. Reproduced by permission of Spark Holland, The Netherlands

Analytical separation and detection: ODS, 25×0.46 cm i.d., gradient mobile phase (acetonitile-aqueous phosphate buffer (pH 7)) at a flow rate of 1 ml min^{-1}. Ultraviolet detection at 249 nm.

Example chromatogram (Figure 4.11) of 300 ml of drinking water sample spiked with 0.1 $\mu g\,l^{-1}$ phenylurea herbicides.

A modified SAMOS system has been reported[13] for the on-line analysis of raw river Tyne water for trace phenolic compounds. Such a system was required because of the high levels of humic substances present in the river. As the humic substances have similar chromatographic and UV characteristics to phenol and cresols, two pre-columns are used for analyte preconcentration. The pre-columns used are both of the type, styrene divinylbenzene copolymer (10×2 mm i.d.). Phenol and both *o*-cresol and *p*-cresol, breakthrough the first pre-column after sampling 52 ml of water at a flow rate of 3 ml min^{-1} and are quantitatively collected on a second pre-column after pumping a further 16 ml of water through the system. The first pre-column is replaced after each analysis while the second pre-column is only replaced during standard routine maintenance. Analytical separation of the phenolic compounds is achieved using a Hypersil BDS-C18 column (250×4 mm i.d.) fitted with a 4×4 mm Lichrospher 100 RP-18 guard column and a gradient mobile phase (acetonitrile-aqueous phosphate buffer). Figure 4.12 shows an example chromatogram for the determination of 1 $\mu g\,l^{-1}$

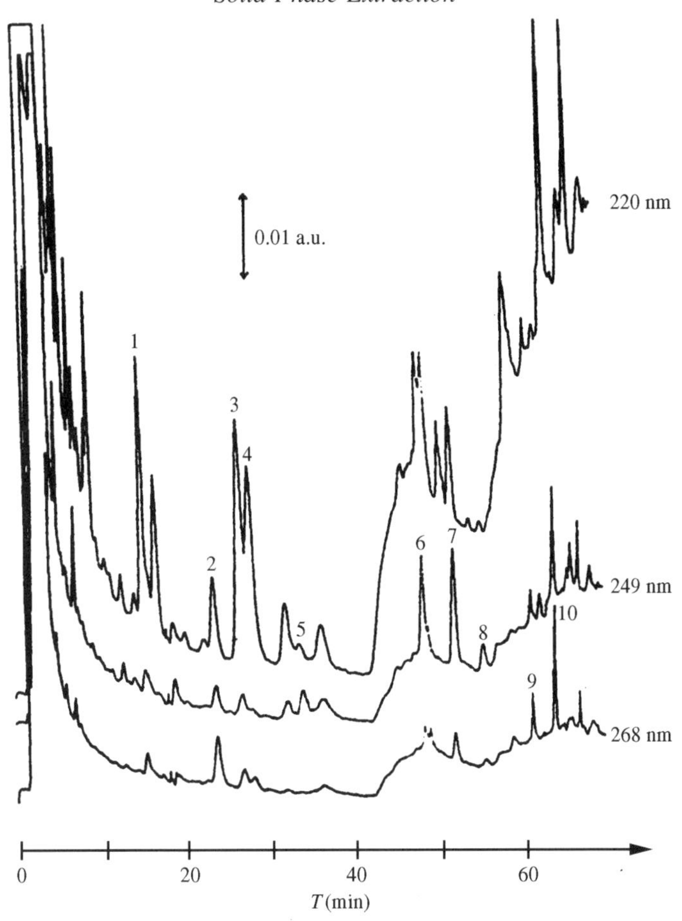

Figure 4.9 Chromatograms corresponding to the on-line elution of 150 ml of drinking water at different wavelengths; drinking water spiked with 0.3 $\mu g\,l^{-1}$ of pesticides. Peak assignment: 1 = simazine; 2 = methabenzthiazuron; 3 = atrazine; 4 = cabaryl; 5 = isoproturon; 6 = propanil; 7 = linuron; 8 = fenamiphos; 9 = fenitrothion; and 10 = parathion. Pre-column, PLRP-S; analytical column, Varian ODS (25 × 0.46 cm i.d.); flow rate, 1 ml min^{-1}; acetonitrile gradient with 0.05 M phosphate buffer at pH 7, gradient 30% acetonitrile from 0 to 38 min, 30 to 45% from 38 to 45 min, 45 to 47% from 44 to 52.5 min, 47 to 100% from 52.5 to 70 min. Reprinted from Pichon and Hennion, *Journal of Chromatography* **665** (1994) 269, with kind permission of Elsevier Science-NL, Sara Burgerhartstraat 25, 1055 KV Amsterdam, The Netherlands

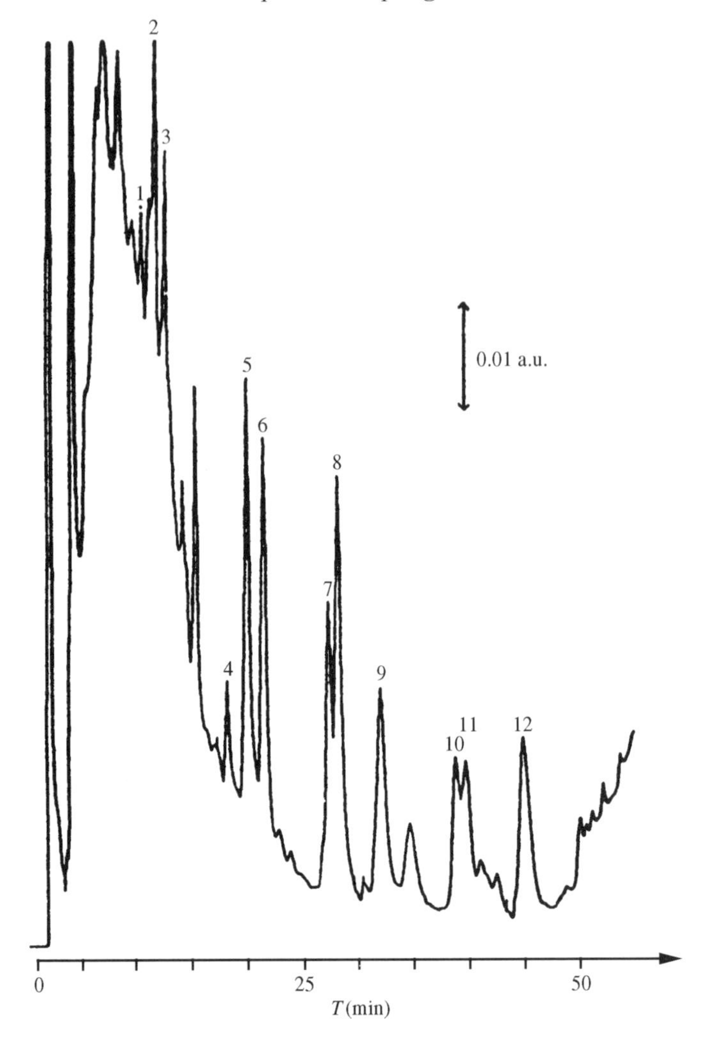

Figure 4.10 Preconcentration on PLRP-S of 300 ml drinking water spiked with 0.1 $\mu g\,l^{-1}$ of triazines. Peak assignment: 1 = desisopropylatrazine; 2 = hydroxyatrazine; 3 = desethylatrazine; 4 = hexazinone; 5 = simazine; 6 = cyanazine; 7 = simetryne; 8 = atrazine; 9 = prometon; 10 = sebuthylazine; 11 = propazine; 12 = terbuthylazine. Analytical column, Varian ODS (25 × 0.46 cm i.d.); flow rate, 1 ml min^{-1}; acetonitrile gradient with 0.05 M phosphate buffer at pH 7, gradient 15 to 30% acetonitrile from 0 to 9 min, 30 to 34% from 9 to 16 min, 34 to 40% from 16 to 45 min, 40 to 60% from 45 to 55 min; detection at 220 nm. Reprinted from Pichon and Hennion *Journal of Chromatography*, **665** (1994) 269, with kind permission of Elsevier Science-NL, Sara Burgerhartstraat 25, 1055 KV Amsterdam, The Netherlands

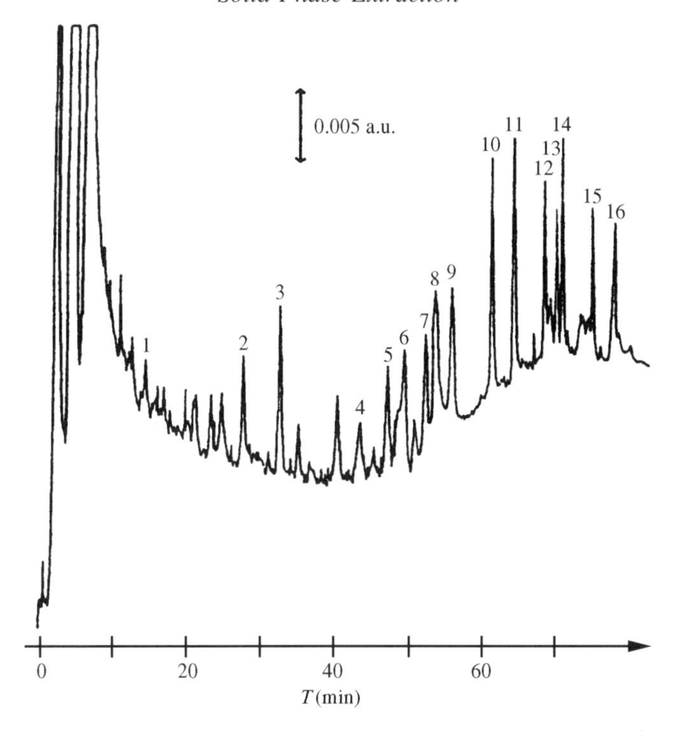

Figure 4.11 Preconcentration on PLRP-S of 300 ml drinking water spiked with 0.1 $\mu g\,l^{-1}$ of phenylureas. Peak assignment: 1 = fenuron; 2 = metoxuron; 3 = monuron; 4 = methabenzthiazuron; 5 = chlorotoluron; 6 = fluometuron; 7 = monolinuron; 8 = isoproturon; 9 = diuron; 10 = difenzoxuron; 11 = buturon; 12 = linuron; 13 = chloroxuron; 14 = chlorbromuron; 15 = diflubenzuron; 16 = neburon. Analytical column, Varian ODS (25 × 0.46 cm i.d.); flow rate, 1 ml min^{-1}; acetonitrile gradient with 0.05 M phosphate buffer at pH 7, gradient 15 to 30% acetonitrile from 0 to 9 min, 30 to 34% from 9 to 16 min, 34 to 40% from 16 to 45 min, 40 to 60% from 45 to 55 min; detection at 220 nm. Reprinted from Pichon and Hennion *Journal of Chromatography*, **665** (1994) 269, with kind permission from Elsevier Science-NL, Sara Burgerhartstraat 25, 1055 KV Amsterdam, The Netherlands

phenols in raw water with (a) a single pre-column and (b) using the preferred two pre-column approach. The authors report excellent long-term repeatability in terms of analyte retention times. In addition, the system is capable of detecting phenol and cresols in the range 0.1–1.0 $\mu g\,l^{-1}$, with confidence, on a 24 hour/day, 7 days/week basis, thereby safeguarding the supply of drinking water to the city of Newcastle upon Tyne and surrounding areas.

The use of ultraviolet-visible detection methods does not offer complete confirmation and/or identification of pollutants, so it is necessary to use mass spectrometry. The availability of lower cost, bench-top HPLC-MS systems enables on-line LC-MS systems to be used.

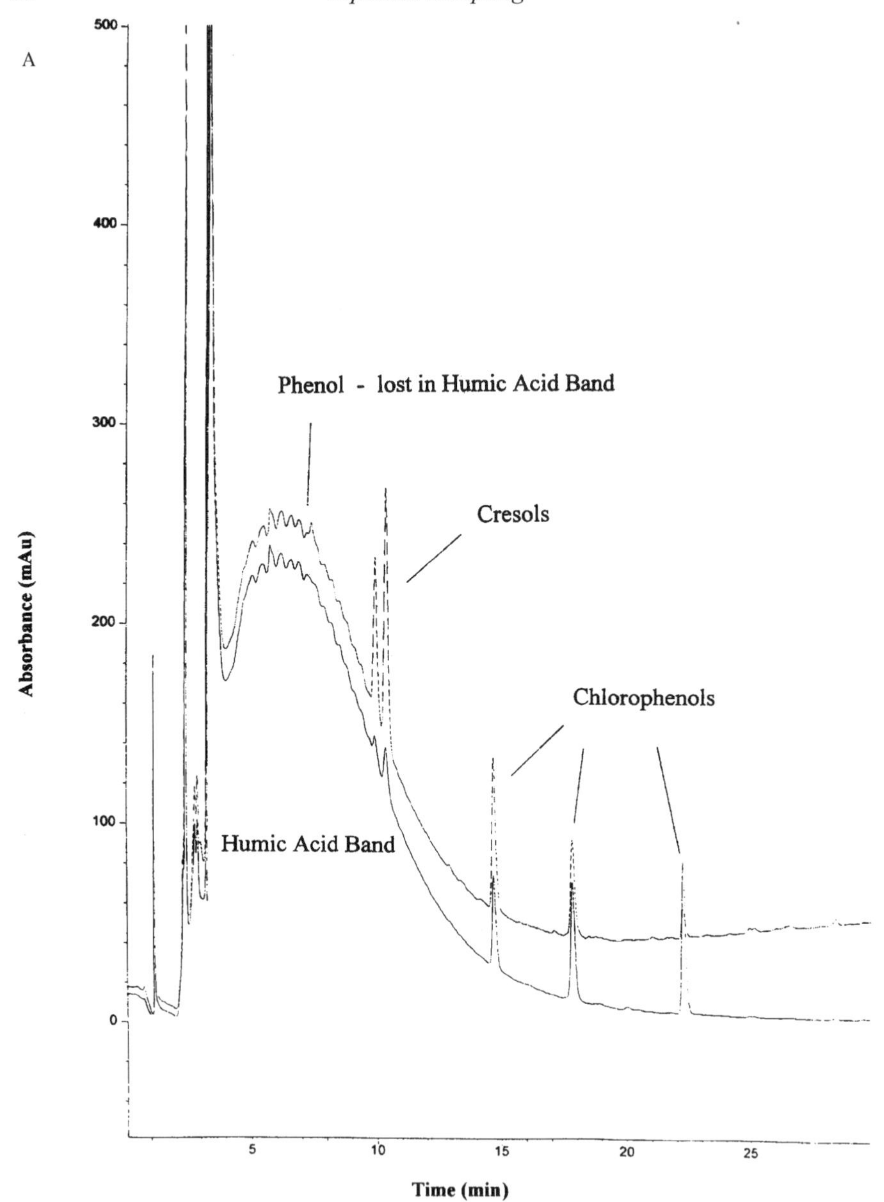

Figure 4.12 Automated SAMOS-HPLC determination of 1 ppb phenols in raw water using two pre-column system. Reproduced by permission of Dr I.A. Fowlis, Northumbrian Water

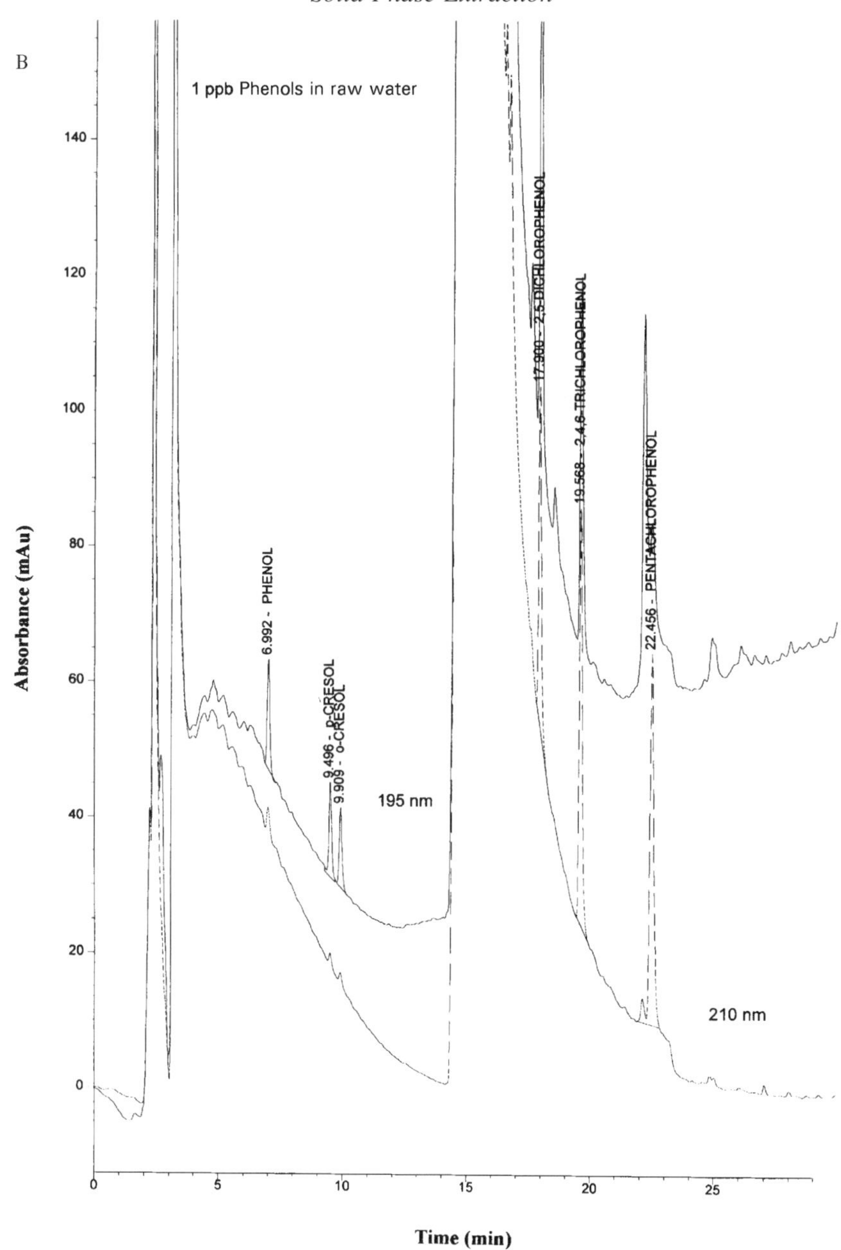

Figure 4.12 (*continued*)

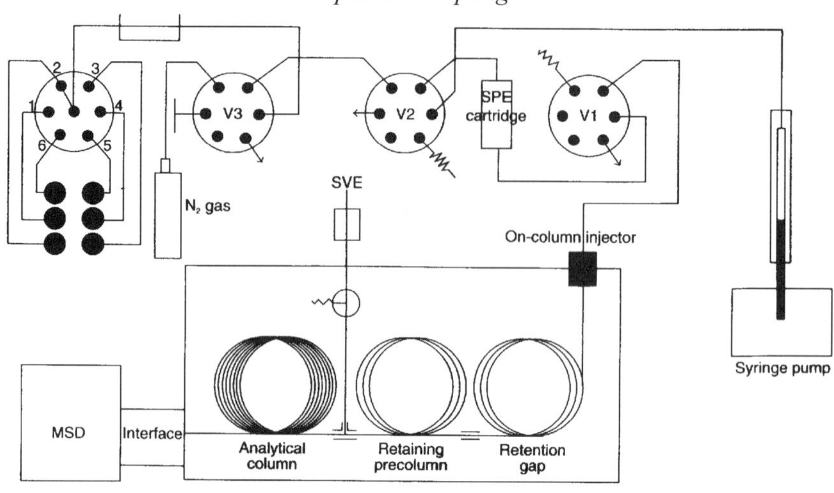

Figure 4.13 Schematic diagram of an on-line SPE-GC-MS system using a solvent selection valve for the conditioning of solvent and samples, and three six-port valves (V1–V3). Nitrogen is used to dry the pre-column (SPE cartridge) after loading of the sample, and ethyl acetate is supplied via a syringe pump for desorption of the analytes into the GC system (retention gap, retaining pre-column and analytical column). SVE = solvent-vapour exit and MSD = mass selective detector. First published in *LC-GC International*, **8** (1995) 694

However, the technique of gas chromatography-mass spectrometry is more commonly available in laboratories. The problem with this experimental arrangement is to combine a 'wet' SPE part of the system with a 'dry' GC system. To combine these two parts requires the development of a suitable SPE-GC interface. Such an interface has been described.[14] Of course, other approaches are available for the introduction of aqueous samples into GC, based on indirect procedures, for example volatile organics in aqueous samples can be analysed using purge and trap or dynamic headspace techniques. For less volatile/more polar compounds the situation is not so clear and this led researchers to investigate the use of on-line SPE. A typical experimental arrangement is shown in Figure 4.13. After preconditioning the pre-column with methanol followed by water, the sample (typically 1–10 ml) is loaded at a flow rate of 1–5 ml min^{-1}. Clean-up is achieved by washing with water, the analytes are desorbed with a small volume (50–100 μl) of an organic solvent (hexane or ethylacetate), at a flow rate of 50 $\mu l\,min^{-1}$, and transferred to the GC part of the instrument for separation and quantitation (Note: Problems can arise due to water in the pre-column, this can be circumvented by passing a stream of nitrogen through the pre-column for approximately 15 min.) It is this transfer that is the most crucial part of the system. One of the most common approaches uses a so-called on-column interface. With this interface, a 3–5 m long retention gap (uncoated, deactivated, fused silica capillary) is used that can be

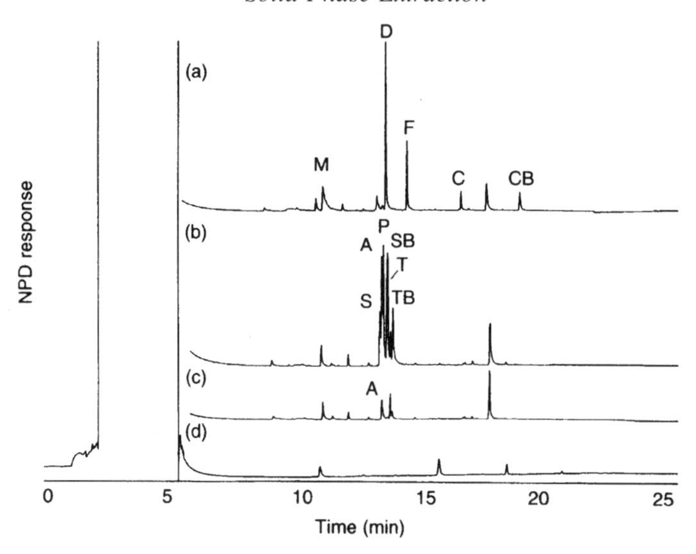

Figure 4.14 SPE-GC-NPD Chromatograms obtained after preconcentration of 10 ml of (d) HPLC-grade water, (c) Amsterdam drinking water, and drinking water spiked with (b) triazine herbicides (0.1 $\mu g\,l^{-1}$) and (a) organophosphorous pesticides (0.3 $\mu g\,l^{-1}$). Peak assignment: M = mevinphos; D = diazinon; F = fenitrothion; C = coumaphos; CB = carbophenthion; S = simazine; A = atrazine; P = propazine; SB = secbumeton; T = trietazine; TB = terbuthylazine. First published in *LC-GC International*, **8** (1994) 694

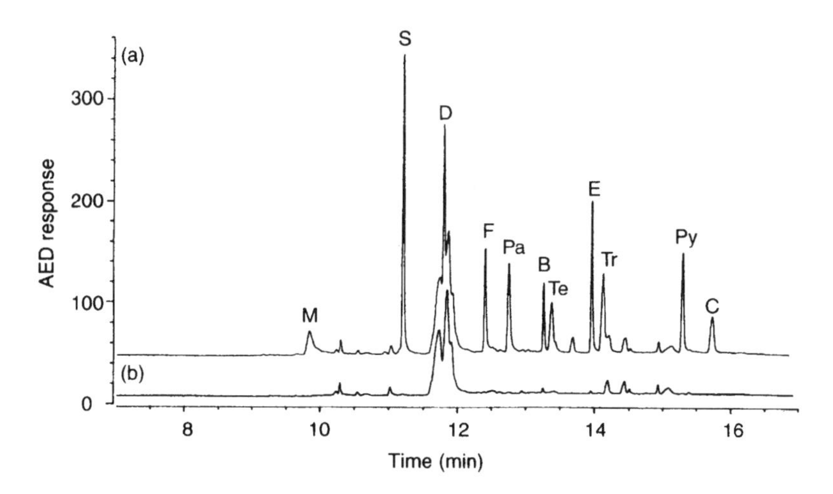

Figure 4.15 On-line SPE-GC-AED (P channel) chromatogram after loading 50 ml of Meuse river water (a) with and (b) without spiking at the 0.1 $\mu g\,l^{-1}$ level. Peak assignment: M = mevinphos; S = sulfotep; D = diazinon; F = fenchlorphos; Pa = parathion; B = bromophos-ethyl; Te = tetrachlorvinphos; E = ethion; Tr = triazophos; Py = pyrazophos; C = coumaphos. First published in *LC-GC International*, **8** (1995) 694

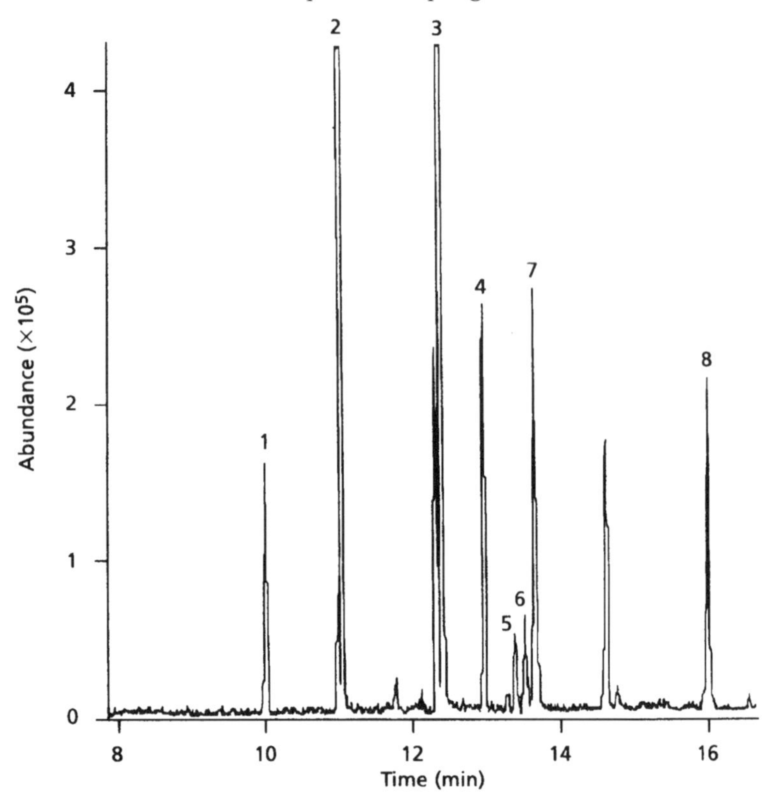

Figure 4.16 On-line SPE-GC-AED chromatogram of triazines in drinking water (45 ppb). Mass spectrometer in SCAN mode. Peak assignment: 1 = C12; 2 = C14; 3 = C16; 4 = trifluralin; 5 = simazine; 6 = atrazine; 7 = propazine; 8 = prometryn. First published in *LC-GC International*, **10** (1997) 435

wetted by the organic solvent used to desorb analytes from the pre-column. Transfer of the analytes occurs below the solvent boiling point (so-called partially concurrent solvent-evaporation conditions). This allows the formation of a solvent film on the inner surface of the retention gap. This process acts to reconcentrate the analytes of interest. At the same time, most of the solvent is rapidly evaporated and vented to waste through the solvent–vapour exit. After completion of this process the exit vent is closed and the GC temperature programme started for the effective separation and subsequent detection of the analytes of interest.

In addition to mass spectrometry, GC also has a diverse range of other detectors available, e.g. flame ionization detection (FID), nitrogen-phosphorus detection (NPD), flame photometric detection (FPD), electron capture detection (ECD) and atomic emission detection (AED). Figure 4.14 shows an example chromatogram of

(a) organophosphorus pesticides, spiked at a concentration of $0.3\,\mu g\,l^{-1}$, in drinking water, and (b) triazine herbicides, spiked at a concentration of $0.1\,\mu g\,l^{-1}$, in drinking water, using a combined SPE-GC-NPD approach. A further example (Figure 4.15), showing the selectivity of the AED detection, monitoring the phosphorus channel, shows the detection of organophosphorus compounds, spiked at a concentration of $0.1\,\mu g\,l^{-1}$, in 50 ml of Meuse river water.

An alternative interface is to use a large volume programmed temperature vaporization (PTV) injector.[15] In this situation, the eluate (100 μl) from the automated SPE system (ASPEC) is introduced directly into the PTV-GC-MS system. Minor instrumental modifications are required to accomplish this experimental set-up. An example chromatogram for the determination of triazines in drinking water ($45\,\mu g\,l^{-1}$) is shown in Figure 4.16.

REFERENCES

1. M. Moors, D.L. Massart and R.D. McDowall, *Pure Appl. Chem.*, **66** (1994) 277.
2. M. Biziuk, J. Namiesnik, J. Czerwinski, D. Gorlo, B. Makuch, W. Janicki, Z. Polkowska and L. Wolska, *J. Chromatogr.*, **733** (1996) 171.
3. E. Viana, M.J. Redondo, G. Font and J.C. Molto, *J. Chromatogr.*, **733** (1996) 267.
4. C. de la Colina, A. Pena, M.D. Mingorance and F. Sanchez Rasero, *J. Chromatogr.*, **733** (1996) 275.
5. M. Terreni, E. Benfenati, M. Pregnolato, A. Bellini, D. Giavini, S.F. Bavetta, C. Molina and D. Barcelo, *J. Chromatogr.*, **754** (1996) 207.
6. M. Honing, J. Riu, D. Barcelo, B.L.M. van Baar and U.A.Th. Brinkman, *J. Chromatogr.*, **733** (1996) 283.
7. G.A. Penuela and D. Barcelo, *J. Chromatogr.*, **754** (1996) 187.
8. M. Vink and J.M. van der Poll, *J. Chromatogr.*, **733** (1996) 361.
9. R. Eisert, K. Levsen and G. Wunsch, *Intern. J. Environ. Anal. Chem.*, **58** (1995) 103.
10. R. Koenig, *Analysis Europa*, March/April (1996) 14, 16.
11. H. Hartman, M. Ettema, C. Klok, H. Kerkdijk, M.H. Godschalk and O. Halmingh, personal communication, Spark Holland.
12. V. Pichon and M.C. Hennion, *J. Chromatogr.*, **665** (1994) 269.
13. I.A. Fowlis, I. Jepson and J.J. Vreuls, *Brighton Crop Protection Conference–Pests and Diseases* (1996), 8B-4, pp. 999.
14. U.A.Th. Brinkman and R.J.J. Vreuls, *LC-GC Int.*, **8** (1995) 694.
15. S. Ollers, M. van Lieshout, H.G. Janssen and C.A. Cramers, *LC-GC Int.*, **10** (1997) 435.

5

Solid Phase Microextraction

Solid phase microextraction or SPME is the process whereby an analyte is adsorbed onto the surface of a coated-silica fibre as a method of concentration. This is followed by desorption of the analytes into a suitable instrument for separation and quantitation. The most important stage of this two-stage process is the adsorption of analyte onto a suitably coated-silica fibre or stationary phase. The choice of sorbent is essential, in that it must have a strong affinity for the target organic compounds, so that preconcentration can occur from either dilute aqueous samples or the gas phase. The range and choice of media available for sorption is ever increasing, and has been exploited most effectively in solid phase extraction (SPE), see Chapter 4. Probably the most reported stationary phase for SPME is poly(dimethylsiloxane). This non-polar phase has been utilised for the extraction of a range of non-polar analytes, e.g. benzene, toluene and xylenes (BTEX) from water and air.[1] The fused silica poly(dimethylsiloxane) coated fibre is stable at high temperatures. This stability and its small physical diameter and cylindrical geometry allow the fibre to be incorporated into a syringe-like holder (Figure 5.1). The SPME holder provides two functions, one is to provide protection for the fibre during transport while the second function is to allow piercing of the rubber septum of the gas chromatograph injector via a needle. As the normal method of introduction of samples into a gas chromatograph is via a syringe the use of a syringe-type device for SPME offers no additional complexity. SPME has been exploited most effectively when coupled to gas chromatography (GC), although more recently it has been used for high performance liquid chromatography (HPLC). In the former case, desorption occurs in the hot injector of the gas chromatograph while the latter relies on the mobile phase for elution. The initial description of SPME will focus on its introduction into GC, as this has been the area initially investigated, and therefore offers the most expansive applications. As we will see later, additional criteria are required when SPME is interfaced to HPLC. The selective nature of the stationary phase of the SPME fibre precludes the introduction of solvent into the gas chromatograph. In addition, no instrument modification is required for the gas chromatograph in terms

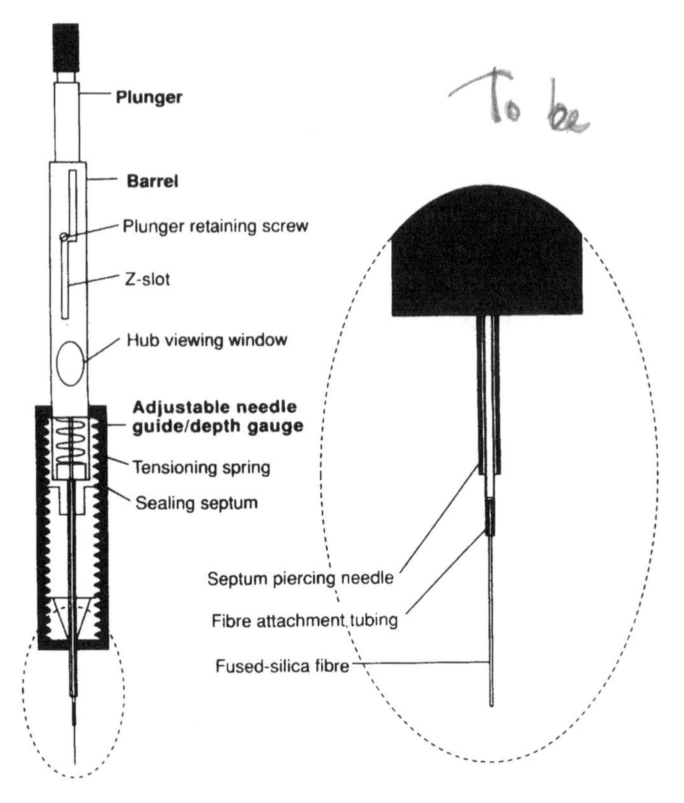

Figure 5.1 Solid phase microextraction device. Reprinted with permission from Zhang *et al. Analytical Chemistry*, **66** (1994) 844A. Copyright (1994) American Chemical Society

of a thermal desorption unit. The heat for desorption from the fibre is provided by the injector of the gas chromatograph.

In the unoperable mode, the fused silica-coated fibre is retracted within the needle of the SPME holder for protection. In operation, however, the coated-silica fibre is exposed to the sample in its matrix. If the sample is liquid full immersion of the coated-silica fibre is required. The active length of the fibre is typically 1 cm. However, it is also possible to extract analytes from the gas phase, e.g. an organic solvent atmosphere in a sealed container (headspace) or the atmosphere in the workplace. In either case, the SPME fibre is exposed to the analyte in its matrix (liquid or gaseous) for a pre-selected time period. After sampling, the fibre is retracted within its holder for protection until inserted in the hot injector of the gas chromatograph. Once located in the hot injector, the fibre is exposed for a particular time to allow for effective desorption of the analytes. As the coating on the fibre is selective towards the analyte, it is common to find that no solvent peaks are present in the subsequent GC trace. Unless precautions are made it is important that the delay between the sorption step and the subsequent desorption and analysis step is small. This is because the silica-coated fibre can equally concentrate analytes from the workplace atmosphere (this might be the sample) as it can from the sample or that losses can occur from the fibre. In the first case the risk of contamination from the workplace environment is high. One way to minimise the risk of contamination for liquid samples at least is to operate the SPME using a modified autosampler on the gas chromatograph. In this case, the sealed vials in the autosampler contain the aqueous samples. In operation, the SPME needle can then pierce an individual vial and carry out the sorption step. This can be immediately followed by insertion into the hot injector of the gas chromatograph. If an automated system is not available contamination from the atmosphere can only be eradicated by minimising the time between extraction and analysis and/or working in a clean room environment. Losses of analyte from the SPME fibre can be achieved by employing some form of preservation. This can be achieved to some extent by cooling the fibre in a fridge or similar.

5.1 THEORETICAL CONSIDERATIONS

The partitioning of analytes between an aqueous sample and a stationary phase is the main principle of operation of SPME. A mathematical relationship for the dynamics of the absorption process was developed by Louch *et al.*[2] In this situation, the amount of analyte absorbed by the silica-coated fibre at equilibrium is directly related to its concentration in the sample, as shown below:

$$n = KV_2C_oV_1/KV_2 + V_1$$

where n = number of moles of the analyte absorbed by the stationary phase; K = partition coefficient of an analyte between the stationary phase and the aqueous phase; C_o = initial concentration of analyte in the aqueous phase; V_1 = volume of the aqueous sample; and V_2 = volume of the stationary phase.

As was stated earlier, the polymeric stationary phases used for SPME have a high affinity for organic molecules, hence the values of K are large. These large values of K lead to good preconcentration of the target analytes in the aqueous sample and a corresponding high sensitivity in terms of the analysis. However, it is unlikely that the values of K are large enough for exhaustive extraction of analytes from the sample. Therefore SPME is an equilibrium method, but provided proper calibration strategies are followed it can provide quantitative data.

Louch *et al.*[2] went on to show that in the case where V_1 is very large (i.e. $V_1 \gg K V_2$) the amount of analyte extracted by the stationary phase could be simplified to:

$$n = KV_2C_o$$

and hence is not related to the sample volume. This feature can be most effectively exploited in field sampling. In this situation, analytes present in natural waters, e.g. lakes and rivers, can be effectively sampled, preconcentrated and then transported back to the laboratory for subsequent analysis.

The dynamics of extraction is controlled by the mass transport of the analytes from the sample to the stationary phase of the silica-coated fibre. The dynamics of the absorption process have been mathematically modelled.[2] In this work, it was assumed that the extraction process is diffusion limited. Therefore, the amount of sample absorbed plotted as a function of time can be derived by solving Fick's second law of diffusion. A plot of the amount of sample absorbed versus time is termed the extraction profile. The dynamics of extraction can be increased by stirring the aqueous sample.

5.2 EXPERIMENTAL

The most common approach for SPME is its use for GC, although as will be seen later its coupling to HPLC has been reported. The SPME device consists of a fused-silica fibre coated with a gas chromatographic stationary phase, e.g. poly-(dimethylsiloxane). In addition, other stationary phases are available for SPME (Table 5.1). The small size and cylindrical geometry allow the fibre to be incorporated into a syringe-type device (Figure 5.1). This allows the SPME device to be effectively used in the normal unmodified injector of a GC. As can be seen in Figure 5.1, the fused-silica fibre (approximately 1 cm) is connected to a stainless steel tube for mechanical strength. This assembly is mounted within the syringe barrel for protection when not in use. For SPME, the fibre is withdrawn into the syringe barrel, then inserted into the sample-containing vial for either solution or air analysis. At this point the fibre is exposed to the analyte(s), by pressing down the plunger, for a prespecified time. After this predetermined time interval the fibre is withdrawn back into its protective syringe barrel and withdrawn from the sample vial. The SPME is then inserted into the hot injector of the GC and the fibre exposed for a prespecified time. The heat of the injector desorbs the analyte(s) from the fibre prior to GC separation and detection. SPME can be done manually or by an autosampler. As the exposed fibre is an active site for adsorption of not only analytes of interest but also air-borne contaminants it is essential that the SPME fibre is placed in the hot injector of the GC prior to adsorption/desorption of analytes of interest to remove potential interferents.

For HPLC analysis using SPME a separate interface is required. The actual adsorption of analytes onto the SPME fibre is the same for both GC and HPLC the difference is the means of desorption. Unlike GC no hot injector is available to desorb the analytes from the fibre. For HPLC therefore, desorption is achieved using the mobile phase of the system. In order to achieve this a separate interface is

Table 5.1 Commonly available SPME fibres

Stationary phase	Thickness (μm)	Maximum exposure temperature (°C)	Recommended desorption temperature (°C)	Comments
Polydimethylsiloxane	100	220	200	High capacity, for volatile, low-mean boiling point (< 220 °C) and apolar compounds, e.g. VOCs
	7	340	220–320	Bound phase for higher desorption temperatures. For semivolatile, high boiling (> 200 °C) and apolar compounds, e.g. PAHs
Polyacrylate	85	310	220–300	High capacity. For both polar and non-polar compounds, e.g. pesticides and phenols

required which is shown in Figure 5.2. The procedure is as follows: before transferring the fibre into the desorption chamber of the interface, the injection valve is placed in the 'load' position. The fibre is then introduced into the desorption chamber by lowering the syringe plunger. The two-piece PEEK union is then closed tightly. The valve is then switched to the 'injection' position, and the desorption procedure started. Solvents from the HPLC pump pass through the desorption chamber in an upstream direction to avoid air bubbles being introduced to the analytical column and disturbing the detector. Analytes that were absorbed by the fibre are then desorbed by the organic solvent and carried to the separation column. Analytical column separation is then initiated and a solvent programme applied to achieve good analytical separation of the analytes of interest.

5.3 METHODS OF ANALYSIS: SPME-GC

5.3.1 VOLATILE ORGANICS IN WATER: DIRECT EXTRACTION

The analysis of benzene, toluene, ethylbenzene and the xylene isomers, the so-called BTEX compounds, has been widely investigated in terms of SPME applications. Potter and Pawliszyn[3] were able to determine a limit of quantitation for benzene from water of 50 pg ml^{-1} with a 100 μm polydimethylsiloxane fibre

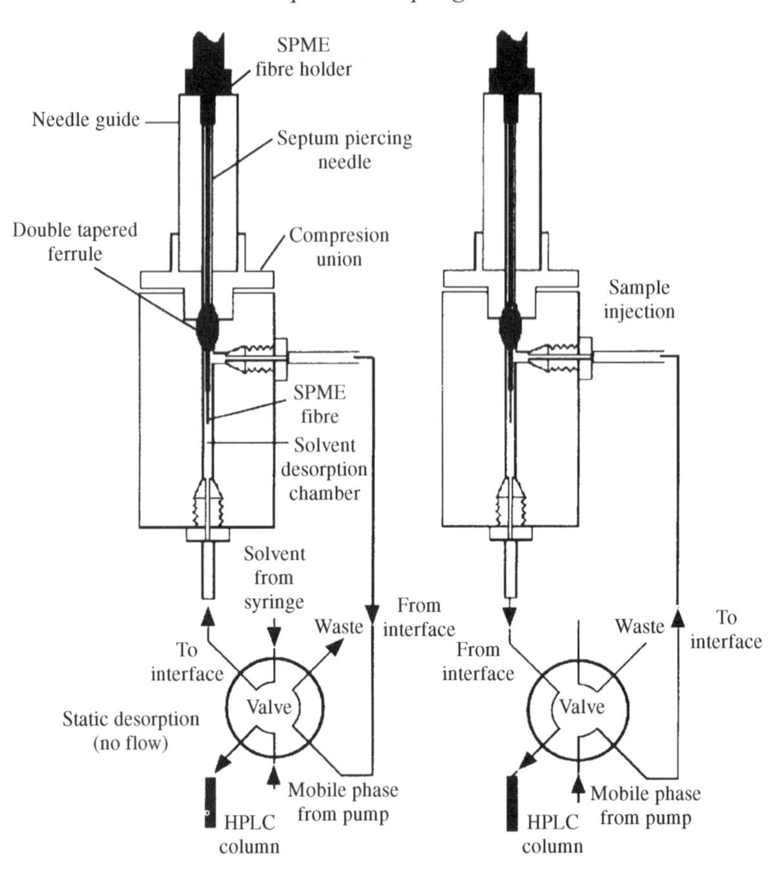

Figure 5.2 SPME-HPLC interface. Reproduced by permission of Supelco UK, Poole, Dorset

coupled with a gas chromatograph-ion trap mass spectrometer. In addition, the linearity of the method extended to over five orders of magnitude. Good precision was obtained using this approach with typical relative standard deviation (RSD) ranging from 2.7 to 5.2% for 15 ng ml^{-1} BTEX in water and from 5.5 to 7.5% for 50 pg ml^{-1} BTEX. The analytical performance of the system is summarised in Table 5.2. The authors used a 30 min adsorption time which was experimentally determined to be double that required to reach equilibriation but convenient as the GC run was of the order of 30 min. Desorption was achieved in a septum programmable injector with a heating programme of 250 °C/min from 25 to 150 °C with a final hold time of 3 min. To focus the analytes the column was cooled to −5 °C for 2 min. No carry-over was noted from the SPME fibre after injection of a 1.5 ng ml^{-1} standard. The combined approach of SPME-GC-ITMS was applied to parking lot run-off water and a coal gasification wastewater sample. A total ion

Table 5.2 Analytical performance criteria obtained using SPME combined with gas chromatography–ion trap mass spectrometry (adapted from Reference 3)

Analyte	Limit of detection ($pg\,ml^{-1}$)	Limit of quantitation ($pg\,ml^{-1}$)	Precision at $50\,pg\,ml^{-1}$ (%)	Precision at $15\,ng\,ml^{-1}$ (%)	Method detection limits as required by the USEPA ($pg\,ml^{-1}$)
Benzene	15	50	7.3	5.3	30
Toluene	5	15	6.7	3.2	80
Ethylbenzene	2	7	7.2	3.6	60
m- and *p*-Xylene	1	4	6.5	6.5	90
o-Xylene	1.5	5	5.5	2.7	60

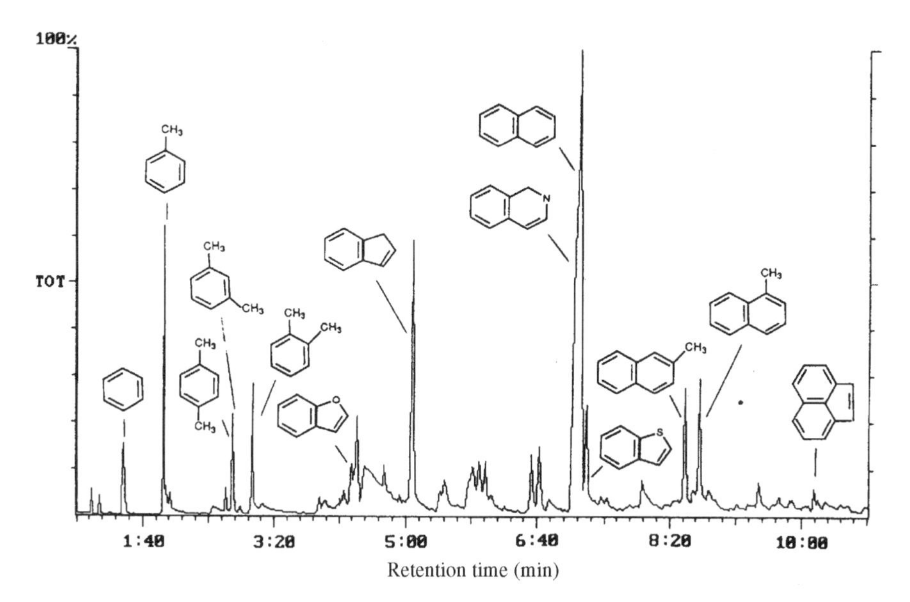

Figure 5.3 Total ion chromatogram showing organics detected and identified in a coal gasification sample. Reprinted from Potter and Pawliszyn, *Journal of Chromatography*, **625** (1992) 247, with kind permission from Elsevier Science-NL, Sara Burgerhartstraat 25, 1055 KV Amsterdam, The Netherlands

chromatogram of the waste water sample is shown in Figure 5.3 (obtained using a 56 μm thick coating and a 5 min adsorption time). A range of aromatics, in addition to benzene and toluene, can clearly be observed, e.g. PAHs. Similar information, from the same group, relating to the determination of BTEX in ground water was also reported.[4] Valor *et al.*[5] compared SPME with headspace analysis for the determination of BTEX in water. The results (Table 5.3) indicate the superior limit of detection achievable using SPME, with the exception of benzene. Benzene has a

Table 5.3 Comparison of SPME with headspace analysis for BTEX in water (adapted from Reference 5)

Analyte	Limit of detection for SPME ($ng\ ml^{-1}$)	Precision of SPME (%) ($n = 5$)	Limit of detection for headspace ($ng\ ml^{-1}$)	Precision of headspace (%) ($n = 5$)
Benzene	0.22	4	0.15	3
Toluene	0.11	3	0.13	8
Ethylbenzene	0.04	9	0.08	6
m- and *p*-Xylene	0.09	9	0.21	6
o-Xylene	0.03	9	0.41	11

low distribution coefficient and a high Henry's constant. The precision of both approaches was similar.

5.3.2 VOLATILE ORGANICS IN WATER: HEADSPACE CONCENTRATION

In addition to placing the SPME fibre directly into the aqueous sample it is possible, provided that the analytes are volatile, to use a headspace approach to SPME. Initial work on headspace SPME was reported by Zhang and Pawliszyn in 1993.[6] In this paper, they report that the sampling time for BTEX in water can be reduced to 1 min compared to direct SPME sampling of the aqueous phase. At ambient temperature, the headspace SPME approach can be applied to compounds with Henry's constants above $90\ atm\ cm^3\ mol^{-1}$ i.e. three-ring PAHs or more volatile species. It was also suggested that the equilibriation time for less volatile compounds can be shortened significantly by agitation of both aqueous phase and headspace, reduction of headspace volume and by increasing the temperature. They also report the feasibility of analysing the headspace above a soil or sewage sample for PAHs.

Headspace SPME was compared with purge and trap for the analysis of BTEX compounds in water.[7] The SPME approach was optimised using a three-factor, two-level statistical design to investigate the effects of temperature, pH and salt addition. It was concluded that the highest sensitivities were achievable at ambient temperature with salt saturation; pH had no effect. Samples in the range 4 to $140\ ng\ ml^{-1}$ were analysed by both techniques and found to be in agreement. Statistical method detection limits, calculated according to the prescribed method of the US Environmental Protection Agency (USEPA), are shown in Table 5.4 and compared with the requirements of EPA methods 624 and 524.2, and the Ontario Municipal and Industrial Strategy for Abatement (MISA) detection limits for effluents. Both headspace and purge and trap were shown to be effective methods of analysis.

Table 5.4 Comparison of headspace SPME and purge and trap for the determination of BTEX in water: statistical method detection limits (ng ml^{-1}) (adapted from Reference 7)

Analyte	Headspace SPME	Purge and trap	EPA method 624	Ontario MISA	EPA method 524.2
Benzene	0.70	0.38	4.4	0.5	0.03
Toluene	0.30	0.37	6.0	0.5	0.05
Ethylbenzene	0.35	0.43	7.2	0.5	0.03
m- and *p*-Xylene	0.23	0.72	NA	1.1	0.05
o-Xylene	0.19	0.30	NA	0.5	0.06

NA = not applicable

5.3.3 PESTICIDES FROM AQUEOUS SAMPLES

The automated determination of four s-triazine herbicides using SPME-GC with a nitrogen-phosphorus flame thermionic detector was reported by Barnabas *et al.*[8] The four triazines studied were simazine, atrazine, propazine and trietazine using either a 7 or 100 μm polydimethylsiloxane coated silica fibre. Initial work compared SPME with a manual GC injection on a 1 ppm solution of the four triazines. It was noted that some selectivity between triazines occurs with respect to SPME. The SPME parameters studied included the effects of adsorption time, desorption temperature, column focusing temperature and dynamic range. It was noted that using the experimental arrangement did not provide enough sensitivity to detect 0.1 ppb of the herbicides in drinking water (EEC limit). In order to circumvent this problem a multiple extraction approach was adopted. Using a 0.1 ppb solution of the herbicides, a 10 min adsorption followed by a 5 min desorption was repeated ten times from a single solution whilst maintaining the GC column at low temperature. This allowed stacking of herbicides on the front of the GC column. After the required number of SPME cycles the GC is operated as normal. The results indicated that this was a suitable approach to detect low concentration solutions by SPME which would normally be well below the limit of detection of the detector used.

A more extensive study of (22) nitrogen-containing herbicides was reported by Boyd-Bowland and Pawliszyn[9] using an 95 μm polyacrylate SPME fibre. A variety of detectors for GC (MS, NPD and FID) were compared for their sensitivity towards the herbicides. Operating parameters for SPME were investigated and include: equilibriation time (ranging from 10–120 min but 50 min chosen for future work); desorption temperature and time (5 min desorption at 230 °C); linearity of the method investigated over the range 0.1–1000 ng ml^{-1}; precision (7–22%RSD, based on $n = 7$ at a concentration of 10 ng ml^{-1}); ionic strength and pH (salt and/or pH 2 investigated and compared with neutral conditions). The calculated limits of detection for the 22 herbicides ranged from 0.01–15 ng l^{-1} using MS detection, 10–6000 ng l^{-1} for the NPD and 200–19000 ng l^{-1} for the FID. The approach was qualitatively applied to the extract of a soil previously treated with a herbicide (Benfluralin).

The on-line determination of 20 organochlorine pesticides (OCPs) in water using SPME and GC with ECD was reported by Young *et al.*[10] A single fibre-type available in three film thicknesses (20, 30 and 100 μm) was evaluated. The extraction time, the effects of stirring and addition of NaCl to the aqueous sample, the linear range and the precision of the technique, and the effects of carry-over were investigated. A comparison of a direct injection and SPME was done (Figure 5.4). It can be seen that selectivity has obviously occurred between the pesticides due to the SPME fibre coating. The SPME method was compared with a liquid–liquid extraction (LLE). The LLE (EPA's Quick-Turnaround Method for OCPs, 12/28/93 draft version) involved extracting a 100 ml sample in a 125 ml Erlenmeyer flask with 20 ml, and then 10 ml hexane using a stirring bar and a magnetic stirrer. The extracts were combined, dried with anhydrous sodium sulphate, and concentrated to 1 ml using a gentle stream of nitrogen. The results, shown in Figure 5.5, indicate that the SPME recoveries were higher than those by LLE, but 11 of the 60 average recoveries for SPME exceeded 120% (in particular endosulfan isomers which had recoveries between 144 and 273%). Excluding the two endosulfan isomers, the recoveries by SPME ranged from 38.8% for 4,4′-DDE to 137% for endosulfan sulphate. For comparison, the average recoveries for LLE ranged from 49.8% for aldrin to 111% for 4,4′-DDT. Precision data was better for LLE than SPME. For LLE only 4 out of 60 values exceeded 10% RSD while for SPME 48 out of 60 values exceeded 10% RSD. The SPME approach was suggested as a fast screening technique for OCPs in water samples.

The simultaneous determination of 60 pesticides in water using SPME and GC-MS was reported by Boyd-Bowland *et al.*[11] The pesticides were chosen to include representatives from each of the following classes (organonitrogen, organochlorine and organophosphorus). Two types of SPME fibre were used i.e. polyacrylate and polydimethylsiloxane, to extract the pesticides over the concentration range 0.1–100 $\mu g\,l^{-1}$. Detection limits for all analytes and using either fibre ranged from 0.1 to 60 $ng\,l^{-1}$. An equilibriation time of 50 min was chosen on the basis of practicality. Typical chromatograms of the 60 pesticides using both types of fibre are shown in Figure 5.6.

The results of an interlaboratory study on the applicability of SPME for the analysis of pesticides in water have been reported.[12] Eleven laboratories (named) in Europe and North America took part in the test. The test sample contained 12 pesticides representing all main groups (organochlorine, organonitrogen and organophosphorus) at low ppb levels. Each laboratory was given a test procedure which included the following steps:

(1) Conditioning of the column according to the manufacturer's specifications and check of the column blank.
(2) Conditioning of the fibre (100 μm polydimethylsiloxane silica coated) according to the manufacturer's specifications and check of the fibre blank.
(3) For quadrupole MS users only: syringe injection of the 10 ppm standard in order to establish the retention times of the analytes and set up the selected-ion monitoring (SIM) acquisition method.
(4) Preparation of a 30 ppb aqueous standard followed by SPME/GC-MS analysis and subsequent carry-over check.
(5) Repeated analysis of a freshly prepared 30 ppb aqueous standard.
(6) Analysis of 10 and 1 ppb aqueous standards in a similar way.
(7) Determination of the calibration curves for all the analytes.
(8) Preparation of the aqueous solution of the blind sample followed by SPME analysis.

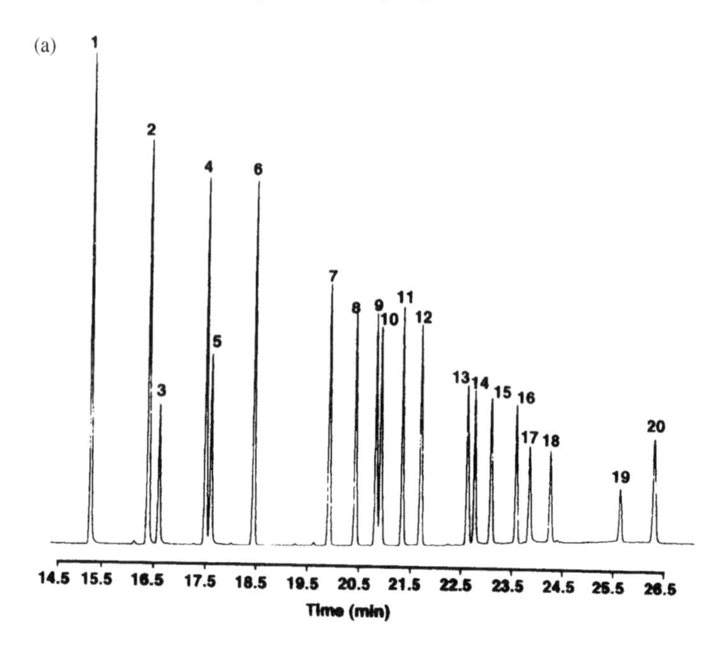
(a)
1
2
3
4
5
6
7
8
9
10
11
12
13
14
15
16
17
18
19
20
14.5 15.5 16.5 17.5 18.5 19.5 20.5 21.5 22.5 23.5 24.5 25.5 26.5
Time (min)

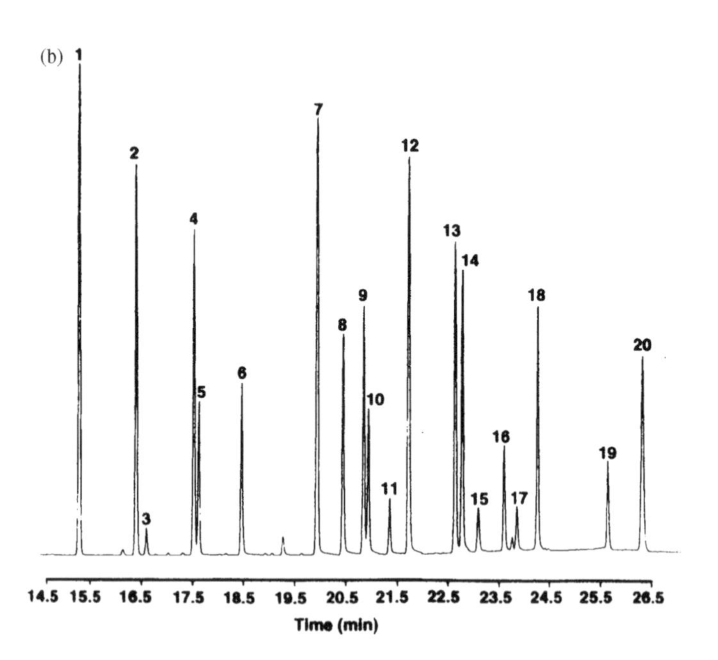
(b)
1
2
3
4
5
6
7
8
9
10
11
12
13
14
15
16
17
18
19
20
14.5 15.5 16.5 17.5 18.5 19.5 20.5 21.5 22.5 23.5 24.5 25.5 26.5
Time (min)

(9) Calculation of the concentration of the analytes in the blind aqueous sample.

The test procedure specified that extractions should be carried out with samples vigorously stirred. The extraction time was set at 45 ± 0.5 min. The chromatographic conditions were as follows: injector and transfer line temperature, 250 °C; temperature programme, 40 °C held for 5 min, increased at 30 °C min^{-1} to 100 °C, at 5 °C min^{-1} to 250 °C and at 50 °C min^{-1} to 300 °C, held for 1 min. Each participant was required to report the results using the forms included in the test kit and to attach the following:

> chromatograms of the column blank and fibre blank; chromatogram of the syringe injection and spectra of the analytes (quadrupole MS only); chromatogram of the 30 ppb standard (all systems) and the spectra of the analytes (ion trap MS only); chromatograms of the remaining standards (one for each level of quantitation); sample chromatogram of the carry-over determination; chromatograms of the unknown sample; and, a copy of the spreadsheet used for the calculation.

The statistical characteristics of the results obtained by the participating laboratories for the blind sample are shown in Table 5.5. In general the results are characterised by good repeatability and satisfactory repeatability. The confidence intervals of the gross average and the 'true' value overlap, which indicates that any differences between the two respective values are due to random factors. It was concluded that SPME is an accurate and fast method of sample preparation and analysis. It can be an excellent alternative to currently used methods.

5.3.4 PHENOLS

A rapid method for the determination of phenols regulated by the USEPA waste water methods 604 and 625 and Ontario MISA Group 20 regulations has been developed using a poly(acrylate) coated SPME fibre.[13] The linear range, detection limit and precision of this approach are summarised in Table 5.6 for a range of phenols. It is seen that the approach is capable of sub ng ml^{-1} limits of detection and precision of 4–12% RSD. Dean *et al.*[14] utilised an automated SPME approach

---

Figure 5.4 (a) Direct injection of 0.25 $\mu g\,l^{-1}$ organochlorine pesticide standard (attenuation 25). (b) SPME of a 1 $\mu g\,l^{-1}$ organochlorine pesticide standard (attenuation 154). GC conditions; 100 °C (4 min hold) to 150 °C at 30 min^{-1} then to 300 °C (8.6 min hold) at 8 °C min^{-1}. Peak assignment: 1 = alpha-BHC; 2 = gamma-BHC; 3 = beta-BHC; 4 = heptachlor; 5 = delta-BHC; 6 = aldrin; 7 = heptachlor epoxide; 8 = gamma-chlordane; 9 = alpha-chlordane; 10 = endosulfan I; 11 = 4,4′-DDE; 12 = dieldrin; 13 = endrin; 14 = 4,4′-DDT; 17 = endrin aldehyde; 18 = endosulfan sulfate; 19 = methoxychlor; 20 = endrin ketone. Reproduced by permission of Hüthig-Fachverlage, from Young *et al.*, *Journal of High Resolution Chromatography*, **19** (1996) 247

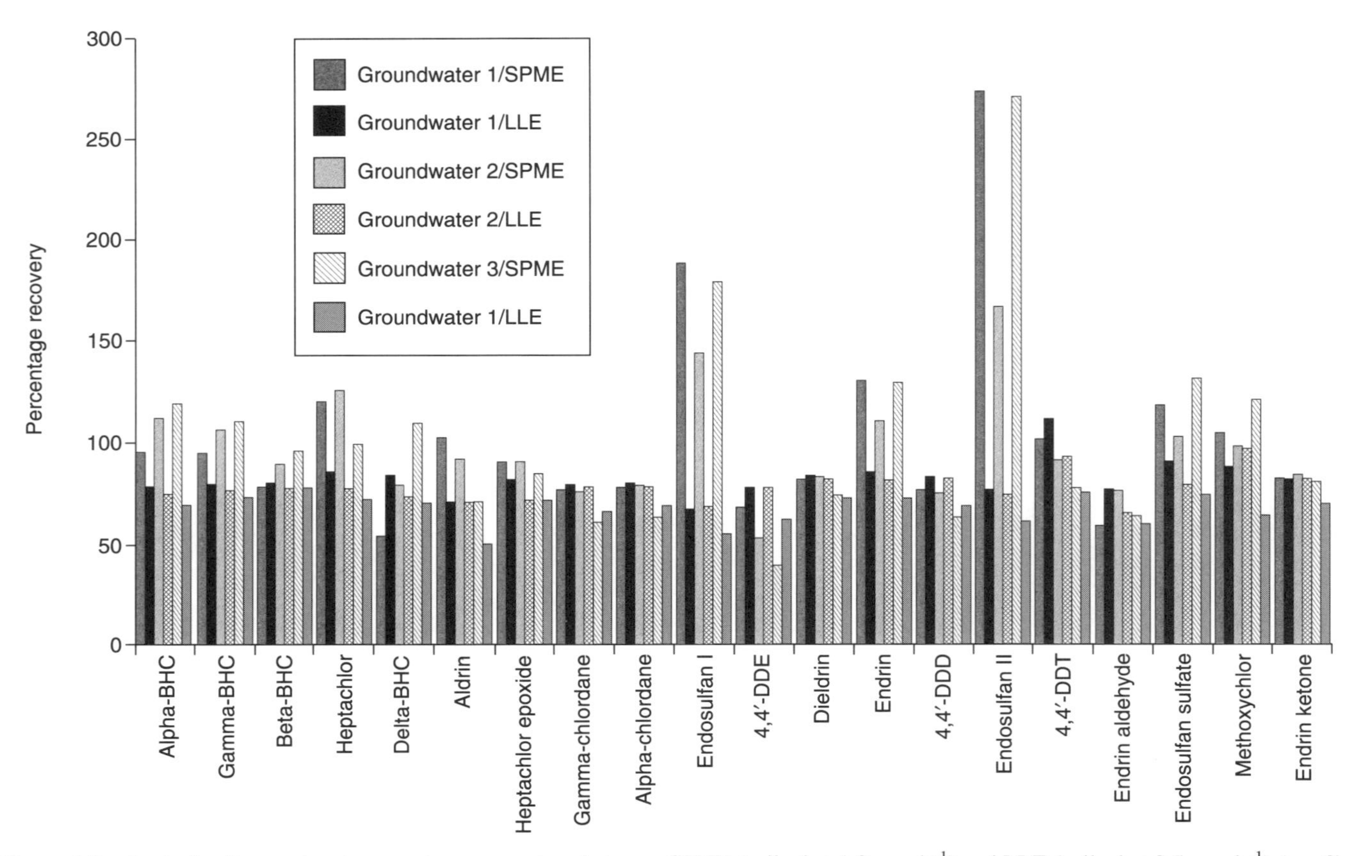

Figure 5.5 Analysis of ground water samples: comparison between SPME (spiked at 1.0 ng ml^{-1}) and LLE (spiked at 0.5 ng ml^{-1}) ($n = 3$). Reproduced by permission of Hüthig-Fachverlage from Young *et al.*, *Journal of High Resolution Chromatography*, **19** (1996) 247

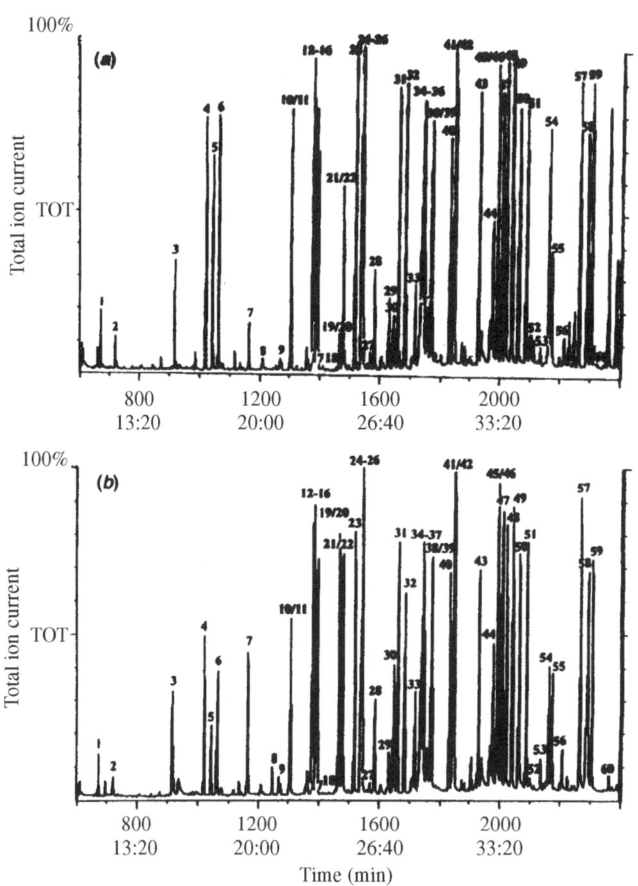

Figure 5.6 Simultaneous determination of 60 pesticides in a mixture by SPME with GC-MS and (a) polydimethylsiloxane and (b) polyacrylate fibres, with a 50 min extraction from $100\,ng\,l^{-1}$ solution. Peak assignment: 1 = O,O,O-TEP; 2 = dichlorovos; 3 = EPTC; 4 = butylate; 5 = vernolate; 6 = pebulate; 7 = molinate; 8 = thionazin; 9 = propachlor; 10 = ethoprofos; 11 = cycloate; 12 = trifluralin; 13 = benxfluralin; 14 = sulfotep; 15 = phorate; 16 = alpha-BHC; 17 = dimethoate; 18 = simazine; 19 = atrazine; 20 = beta-BHC; 21 = propazine; 22 = lindane; 23 = profluralin; 24 = diazinon; 25 = delta-BHC; 26 = disulfoton; 27 = terbacil; 28 = iprobenfos; 29 = metribuzin; 30 = methyl parathion; 31 = heptochlor, 32 = fenchlorovos; 33 = fenitrothion; 34 = bromacil; 35 = isoxathion; 36 = aldrin; 37 = metochlor; 38 = chlorpyrifos; 39 = ethyl parathion; 40 = isopropalin; 41 = pendimethalin; 42 = heptachlor; 43 = epoxide; 44 = prothiofos; 45 = p,p′-DDE; 46 = dieldrin; 47 = oxadiazon; 48 = oxyflurofen; 49 = endrin; 50 = endosulfan II; 51 = p,p′-DDD; 52 = endrin aldehyde; 53 = famphur; 54 = endosulfan sulfate; 55 = p,p′-DDT; 56 = hexazinone; 57 = endrin ketone; 58 = EPN; 59 = methoxychlor; 60 = azinphos-methyl. Reproduced by permission of the Royal Society of Chemistry from Boyd-Bowland *et al.*, *Analyst*, **121** (1996) 929

Table 5.5 Statistical characteristics of the results obtained by the participating laboratories for the blind sample[12]

Compound	S_r	S_L	S_R	r	R	GA	CI	TV
Dichlorvos	2.06	5.04	5.44	5.83	15.40	27.3	27 ± 5.8	25 ± 1.35
EPTC	0.56	1.56	1.66	1.57	4.70	9.9	10 ± 1.6	10 ± 0.54
Ethoprofos	0.82	4.79	4.86	2.32	13.74	15.5	16 ± 2.3	17 ± 0.92
Trifluralin	0.27	0.57	0.63	0.76	1.79	1.6	1.6 ± 0.76	2 ± 0.11
Simazine	2.34	3.45	4.17	6.61	11.79	23.6	24 ± 6.6	25 ± 1.35
Propazine	1.21	2.04	2.37	3.42	6.71	9.5	10 ± 3.4	10 ± 0.54
Diazinon	0.63	2.13	2.22	1.79	6.29	8.2	8 ± 1.8	10 ± 0.54
Methyl chlorpyriphos	0.12	0.32	0.34	0.35	0.97	1.6	1.6 ± 0.35	2 ± 0.11
Heptachlor	2.03	2.89	3.53	5.75	10.00	8.9	9 ± 5.8	10 ± 0.54
Aldrin	0.54	0.73	0.91	1.53	2.58	2.0	2 ± 1.5	2 ± 0.11
Metolachlor	0.73	2.83	2.92	2.07	8.28	15.7	16 ± 2.1	17 ± 0.92
Endrin	0.87	3.00	3.13	2.47	8.85	8.8	9 ± 2.5	10 ± 0.54

S_r = repeatability standard deviation; S_L = interlaboratory standard deviation; S_R = reproducibility standard deviation; r = repeatability; R = reproducibility; GA = gross average; CI = confidence interval of the gross; TV = confidence interval of the 'true' value. All values expressed in $\mu g\,l^{-1}$.

for estimating the octanol-water partition coefficient (log K_{ow}) for a range of phenols (six). Good agreement was obtained between the SPME approach and log K_{ow} values obtained by calculation or experimentally determined. The time scale to reach equilibriation was prolonged (up to 20 hours), however, it was suggested that this could be reduced by stirring of the solution or agitation of the fibre.

5.3.5 ANALYSIS OF ANALYTES FROM SOLID MATRICES

The use of SPME to quantify the level of pollutants in soil slurries has been presented by several authors.[9,11,15–17] The intention is that a known quantity of soil is stirred with water and then to expose the SPME fibre directly to the resultant slurry prior to analysis. An initial attempt to demonstrate this application was presented by Boyd-Bowland and Pawliszyn.[9] In this paper a qualitative experiment was done using a soil-water slurry to identify the presence of Benfluralin on a sample of lawn to which the herbicide had been applied. The GC-MS chromatogram indicated the presence of the herbicide. Further work by the same group was done[11] for the assay of metolachlor, a pesticide, on soil. The soil was analysed by stirring 0.5 g in 4 ml of water and then exposing the SPME fibre directly to the resultant slurry for 50 min. The concentration of metolachlor was determined to be 1.84 mg kg^{-1}, which was in agreement with the result obtained by Soxhlet extraction (1.85 mg kg^{-1}). This indicated that for metolachlor, a relatively water-soluble compound (solubility = 530 mg l^{-1}), extraction from soil by the

Table 5.6 Analytical figures of merit for the determination of phenols by SPME using a poly(acrylate) coated fibre (adapted from reference 13)

	SPME linear range ($\mu g\,ml^{-1}$)		SPME detection limits ($ng\,ml^{-1}$)		EPA detection limits ($ng\,ml^{-1}$)		Precision of SPME
Compounds	GC-FID	GC-MS	GC-FID	GC-MS	GC-FID	GC-MS	(% RSD)
Phenol	0.2–2.0	0.007–0.7	30	0.80	0.14	1.5	4.2
2-Chlorophenol	0.02–2.0	0.007–0.7	0.61	0.24	0.31	3.3	4.2
2-Nitrophenol	0.02–2.0	0.007–0.7	11	0.38	0.45	3.6	5.2
2,4-Dimethylphenol	0.02–2.0	0.007–0.7	2.1	0.02	0.32	2.7	4.8
2,4-Dichlorophenol	0.002–2.0	0.007–0.7	0.64	0.02	0.39	2.7	4.9
4-Chloro-3-methylphenol	0.008–8.0	0.007–0.7	1.4	0.01	0.36	3.0	4.0
2,4,6-Trichlorophenol	0.05–5.0	0.007–0.7	0.80	0.08	0.64	2.7	4.5
2,4-Dinitrophenol	0.5–5.0	0.07–0.7	32	1.6	13.0	42.0	8.9
4-Nitrophenol	0.08–8.0	0.007–0.7	7.8	0.75	2.8	2.4	9.3
2-Methyl-4,6-dinitrophenol	0.8–8.0	0.007–0.7	1.7	0.44	16.0	24.0	5.6
Pentachlorophenol	0.008–8.0	0.007–0.7	1.4	0.11	7.4	3.6	12.0

addition of water is as effective as Soxhlet extraction. Unfortunately this is not likely to be the case for other compounds.

This approach has been extended to semivolatile organics including polycyclic aromatic hydrocarbons (PAHs) by Hageman *et al.*[15] In this work hot water (250 °C) is used to extract soil samples. After 15–60 min extraction, the cell is cooled and the water removed from the extraction cell and the solubilised organic analytes are analysed using SPME. Recoveries for PAHs at 250 °C and 15 min range from 45 to 197% and for 60 min from 61 to 278% for railroad bed soil (compared to Soxhlet extraction), based on 11 PAHs. While recoveries for PAHs at 250 °C and 15 min range from 33 to 149% and for 60 min from 58 to 141% for urban dust (SRM 1649), based on eight PAHs (or grouped isomers). The same paper also includes results for a range of volatile and polar organics extracted from an industrial soil using hot water (at 250 °C) and a 15 min SPME. The results are compared with sonication and Soxhlet and are shown in Figure 5.7.

Further work in this area has been presented by Daimon and Pawliszyn[16] who coupled SPME with high temperature water extraction for the determination of PAHs in solid matrices. Two different SPME approaches were evaluated, dynamic and static, using a 30 μm poly(dimethylsiloxane) coated fibre. In dynamic

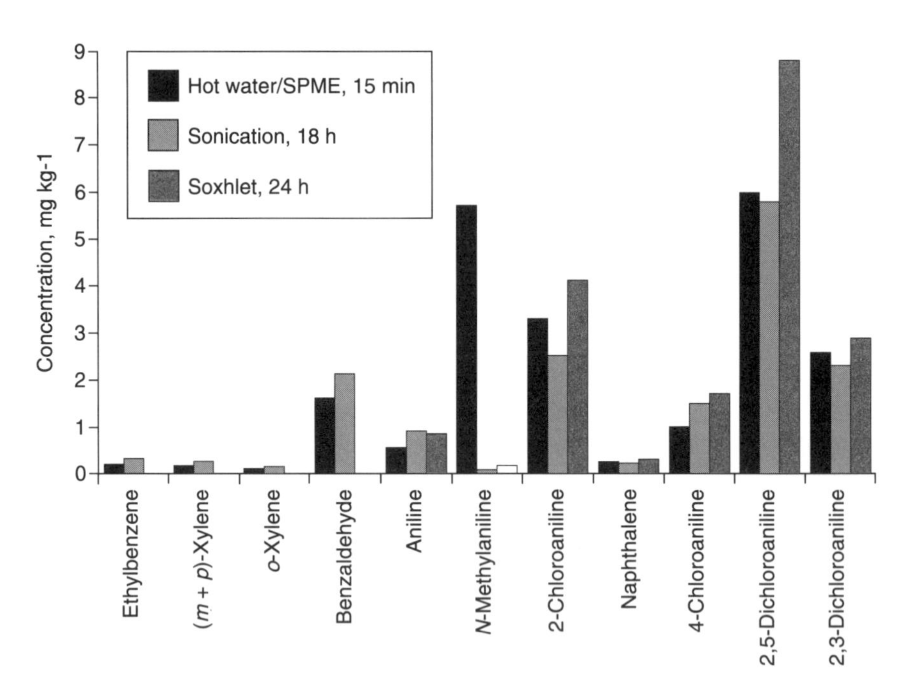

Figure 5.7 Determination of volatile and polar analytes in an industrial soil using hot water extraction (250 °C) followed by SPME. From Hageman *et al.*, *Analytical Chemistry*, **68** (1996) 3892

(a)

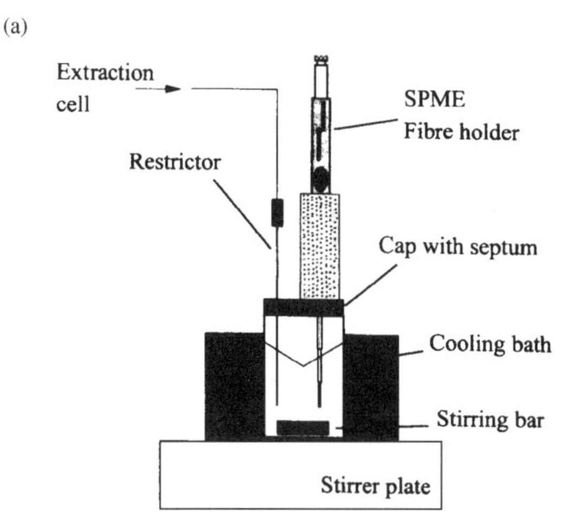

(b)

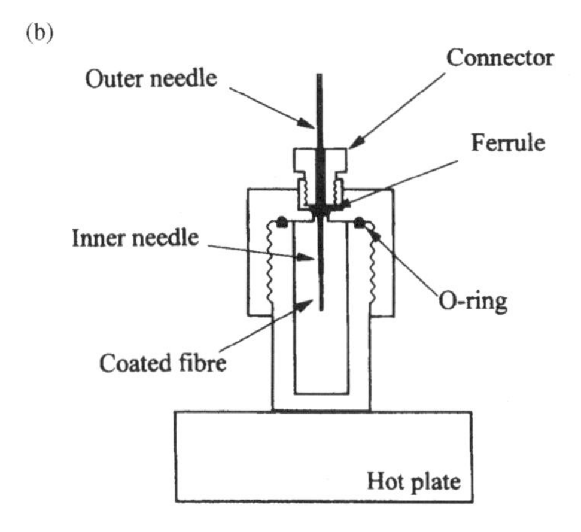

Figure 5.8 (a) Schematic diagram of apparatus for collection method after dynamic high temperature water extraction; (b) schematic diagram of high temperature extraction cell for static high temperature water extraction with simultaneous SPME. Reproduced by permission of the Royal Society of Chemistry from Daimon and Pawliszyn, *Analytical Communications*, **33** (1996) 421

extraction, analytes leached from the matrix with hot water are collected in a vial and simultaneously extracted from the water using SPME. While in the static approach a high pressure cell is used with an SPME fibre inserted in the vessel. During the extraction process analytes released from the solid matrix are partitioned into the fibre coating as the cell is cooling. The experimental arrangement of both systems is shown in Figure 5.8(a) and (b). After SPME the fibre is removed from either the collection water or the high pressure cell and

Table 5.7 Determination of PAHs in urban air particulate (NIST SRM 1649) with static high temperature water extraction and simultaneous extraction with SPME[16]

Compounds	Certificate concentration, [$\mu g\ g^{-1}$ (%RSD)]	Estimated concentrations as percentage recoveries of certified concentration [$\mu g\ g^{-1}$/(% RSD)][a]		
		100 mg sample and 270 °C	50 mg sample and 270 °C	50 mg sample and 300 °C
Fluoranthene	7.1 (7)	137 (6)	149 (8)	128 (2)
Pyrene	7.2 (7)	113 (9)	121 (7)	105 (7)
Benzo[*a*]pyrene	2.9 (17)	49 (14)	72 (10)	70 (4)

[a] $n = 3$

immediately transferred to the injector of the gas chromatograph without any clean-up or preconcentration. Both static and dynamic extraction were evaluated with respect to the analysis of PAHs from urban air particulate (NIST SRM 1649). In the dynamic approach, extraction was done for 15 min at 250 °C and 50 atm. For static extraction the sample was heated to either 270 or 300 °C for 120 min in the presence of water and SPME was done for 120 min while the cell was cooling. The results for the dynamic approach are shown in Figure 5.9 while Table 5.7 contains the results for the static extraction. In both cases acceptable results were obtained using the hot water-SPME method. It was suggested that the advantage of the dynamic approach was the simple external calibration and fibre protection, since the fibre coating is exposed to relatively pure water under low temperature conditions. On the other hand, static high temperature water extraction with simultaneous SPME uses only simple apparatus and procedure, and extends the approach for analytes tightly bound to soils.

An alternative strategy to solid analysis is to use SPME to extract analytes from the headspace above a sample. In this approach James and Stack[17] exposed a 100 μm poly(dimethylsiloxane) fibre to the headspace above a soil sample (1.0 g) heated in the range 22–60 °C. After 180 min the fibre was inserted into the injector of a gas chromatograph and analysed for a selected range of volatile organic compounds. Calibrations, using soil samples spiked with selected solvents (0.5–30 $\mu g\ g^{-1}$), were linear; trichloroethene ($r^2 = 0.992$) and benzene ($r^2 = 0.998$). This approach (50 °C for 30 min) was applied to eight sites within a municipal landfill, at three depths, and resulted in the detection of xylene (maximum 2.8 $\mu g\ g^{-1}$, using the method of standard additions) and a number of other non-target organic contaminants. This application for SPME provided a rapid protocol for the quantification of residual solvents in landfill soil samples.

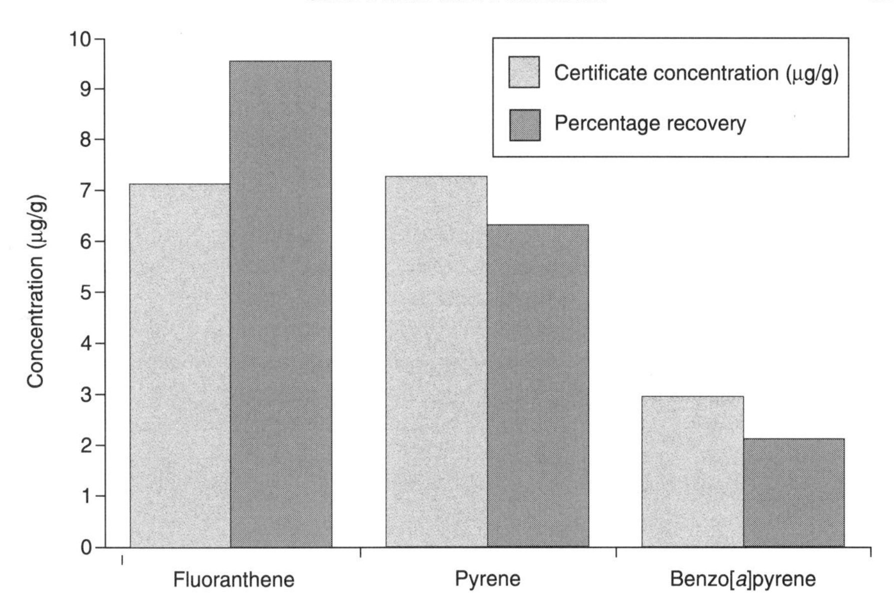

Figure 5.9 Determination of PAHs in urban air particulate (NIST SRM 1649) with dynamic high temperature (250 °C and 50 atm) water extraction and SPME. Reproduced by permission of the Royal Society of Chemistry from Daimon and Pawliszyn, *Analytical Communications*, **33** (1996) 421

5.4 METHODS OF ANALYSIS: SPME-HPLC

It was perhaps logical to assume that after the initial development of SPME for gas chromatography (GC) attention would also focus on the use of SPME with high performance liquid chromatography (HPLC). However, unlike in GC where the injector provides the means for thermal desorption of analytes from the fibre no such situation exists for HPLC. For HPLC therefore, analytes are desorbed from the fibre using the mobile phase, i.e. solvent desorption. This required the development of a separate interface, shown schematically in Figure 5.2. The scientific literature in this area is more limited probably as it is (a) a relatively new approach and (b) requires the use of an interface. Initial work, reported by Chen and Pawliszyn[18] focused on the interfacing of SPME with HPLC using the separation and identification of PAHs as an example. In this work a 7 μm poly(dimethylsiloxane) fibre was exposed to a stirred water sample spiked with PAHs for 30 min. A comparison between a direct 1 μl loop injection and a fibre injection using 7 μm poly(dimethylsiloxane) extraction for 30 min from a 100 ppb solution of each PAH is shown in Figure 5.10. It is noticeable that some fibre selectivity has occurred for the first four peaks separated, i.e. acenaphthylene, fluorene, phenanthrene and anthracene. The same approach has also been applied to the separation of alkylphenol ethoxylate surfactants in water.[19] In this paper normal-phase gradient elution with detection at 220 nm was applied for the analysis of Triton X-100 and other alkyphenol ethoxylates (nonylphenol ethoxylates) after SPME extraction. As well as the commercial fibres, poly(acrylate) and poly(dimethylsiloxane), several experimental fibre coatings were evaluated, namely, Carbowax/template resin, Carbowax/divinylbenzene, poly(dimethylsiloxane)/template resin, poly(dimethylsiloxane)/divinyl benzene and β-cyclodextrin. In addition, extraction time, desorption time and addition of salt were investigated. The fibre that produced the best agreement between the distribution of ethoxamers before and after extraction was the Carbowax/template resin. The linear range of this fibre was determined to be 100–0.1 mg l^{-1}, with limits of detection for individual alkylphenol ethoxylates at the low ppb level. Various application notes on the development of SPME-HPLC can be found in *The Reporter* produced by Supelco.

5.5 MISCELLANEOUS APPLICATIONS

5.5.1 CHARACTERISATION OF ALCOHOLIC BEVERAGES

The extraction, identification and quantification of wine aroma compounds was reported by De la Calle Garcia *et al.*[20] The aim of the work was the systematic optimisation of SPME for analysis of a class of wine aroma compounds, the monoterpenes, which form an important part of the grape bouquet. The monoterpenes investigated were linalool, citronellol, nerol, geraniol, α-terpineol and nerol oxide. Analysis was done using an 85 μm polyacrylate coated silica fibre and 1.3 ml samples. The parameters varied were: time and speed of agitation, volume, temperature and matrix conditions of the sample, i.e. salt concentration, pH and ethanol content. The optimum SPME extraction conditions were identified as follows: time of immersion of fibre in sample, 15 min; speed of agitation, maximum (1500 min^{-1}); temperature, 60 °C; salt concentration, saturated (25% NaCl); and a pH, 4. In addition, injection must be done at 300 °C and the fibre must remain in the middle of the injector for 5 min. The method of standard additions was used for quantitation. Detection limits of the order of 0.1 to 0.5 $\mu g\, l^{-1}$ were obtained for the monoterpenes, suitable for analysis of wine. An example chromatogram is shown in Figure 5.11.

A reliable method for the characterisation of vodkas has been developed using SPME and GC-MS.[21] Fifty-three samples of 18 brands of commercial Canadian vodkas, one sample each of 11 brands of commercial American vodkas, and one sample each of Japanese and German vodka were analysed. The SPME fibre used was a 100 μm polydimethylsiloxane coated silica fibre. Vodka samples (100 ml) were analysed using an internal standard, methyl hexadecanoate (prepared in absolute ethanol). 20 ml sub-samples were equilibriated for 1 hour (not optimum) at room temperature with magnetic stirring before insertion into the GC-MS. The

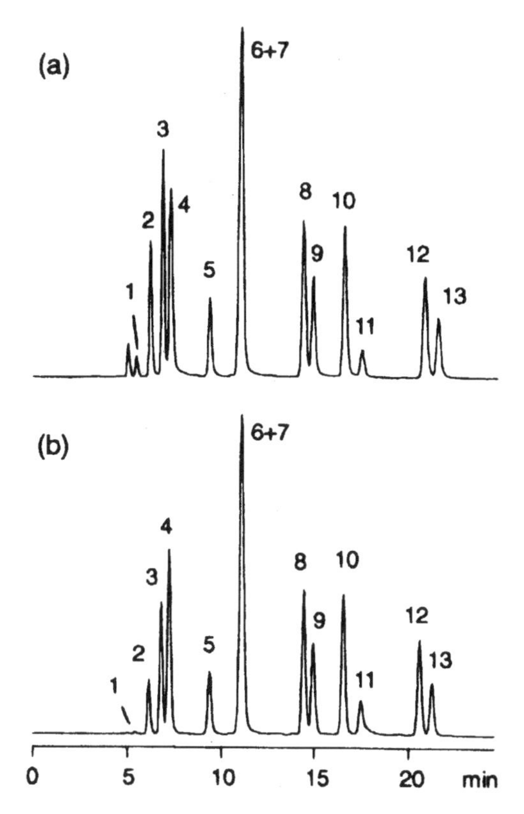

Figure 5.10 Separation of PAH mix with solvent gradient by (a) 1 μl loop injection and (b) fibre injection, 7 μm PDMS extraction for 30 min from 100 ppb of each compound spiked into water. Peak assignment: 1 = acenaphthylene; 2 = fluorene; 3 = phenanthrene; 4 = anthracene; 5 = pyrene; 6 = benz[*a*]anthracene; 7 = chrysene; 8 = benzo[*b*]fluoranthene; 9 = benzo[*k*]fluoranthene; 10 = benzo[*a*]pyrene; 11 = dibenzo[*ah*]anthracene; 12 = indeno[1,2,3-*cd*]pyrene; 13 = benzo[*ghi*]perylene. Chromatographic conditions: column, 25 cm × 2.1 mm i.d., 5 μm ODS; flow rate 0.2 ml min^{-1}; detection, UV 254 nm; solvent program, acetonitrile/water (80/20 v/v) linear gradient to 100% acetonitrile in 15 min. Reprinted with permission from Chen and Pawliszyn, *Analytical Chemistry*, **67** (1995) 2530. Copyright (1995) American Chemical Society

main detected components were ethyl esters of C8 to C18 fatty acids at the μg l^{-1} level but other components were detected in some vodkas, e.g. di(2-ethylhexyl)-phthalate (DEHP), di(2-ethylhexyl)adipate (DEHA), 2,6-di-*tert*-butyl-4-methylphenol (BHT), 5-hydroxymethyl-2-furaldehyde (5-HMF), triethyl citrate (TEC), terpenes and sesquiterpenes. (Note: DEHP and DEHA are common plasticisers; BHT an antioxidant; 5-HMF and TEC are commonly added to vodkas.) It was suggested that the presence of 5-HMF and TEC, predominantly in American

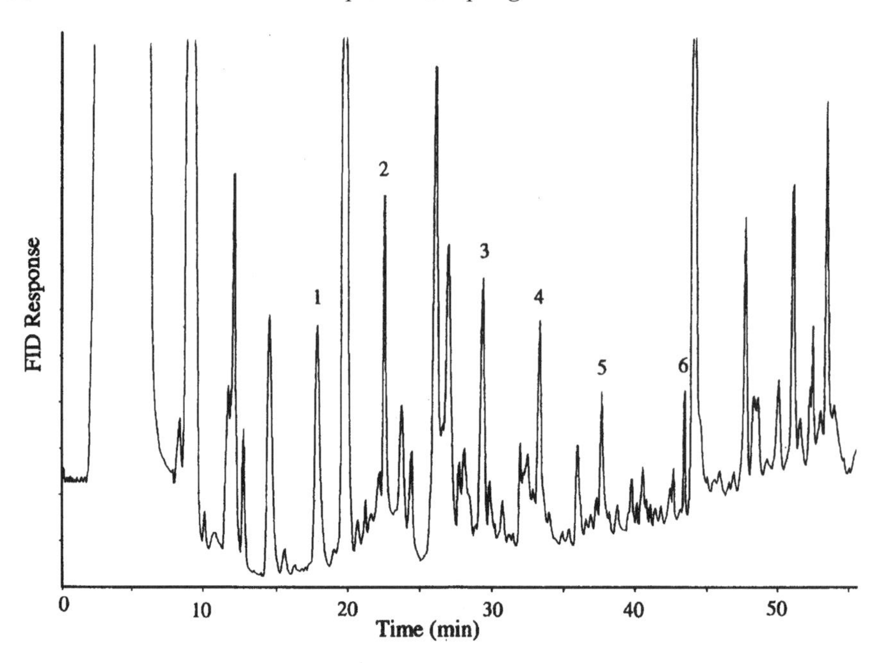

Figure 5.11 Typical chromatogram obtained from wine (Muller-Thurgau, 1993, Freiburg-Unstrut, Germany). Peak assignment: 1 = nerol oxide; 2 = linalool; 3 = alpha-terpineol; 4 = citronellol; 5 = nerol; 6 = geraniol. Reproduced by permission of Hüthig-Fachverlage from Garcia *et al.*, *Journal of High Resolution Chromatography*, **19** (1996) 257

vodkas, was due to the addition of sugars and citric acid, respectively. The authors report that the chromatograms obtained allow distinction between various brands of vodka.

5.5.2 ANALYSIS OF HUMAN BREATH

The determination of ethanol, acetone and isoprene in human breath has been reported using a modified SPME holder with quantitation by GC-MS.[22] Breath monitoring can be used as a diagnostic tool as increased or decreased concentrations of some compounds have been associated with various diseases or altered metabolism.[23] The authors[22] cite information where the use of breath monitoring may be important as it has the advantage of being non-invasive and hence does not require the collection of blood or urine samples. For example, increased concentrations of acetone ($> 50\,\mathrm{nmol\,l^{-1}}$) have been detected in the breath of patients suffering from diabetes. Also the monitoring of acetone levels in breath for patients on diets may serve as a motivational tool during weight-loss programmes. A weight reduction of approximately 225 g week^{-1} is indicated by an acetone concentration of $500\,\text{nmol}\,\text{l}^{-1}$. Eating meals with a high carbohydrate content significantly decreases this concentration. For example ethanol detection on breath is used as it directly relates to the concentration in blood. A breath-to-blood ratio of 1 : 2100 is used when breath alcohol measurement devices are calibrated to reflect blood alcohol concentrations.

Four different SPME fibres were used for this work[22]: a 100 μm polydimethylsiloxane fibre; an 85 μm polyacrylate fibre; a 65 μm polydimethylsiloxane/divinylbenzene fibre; and a 65 μm Carbowax/divinylbenzene fibre. Each fibre was assessed on the basis of sensitivity and extraction time profile for the target analytes. Details of the preparation of gaseous standards are included in the paper. An investigation into the factors that influence SPME are reported, i.e. extraction time, fibre coating, linear range, precision, accuracy, extraction temperature, relative humidity, matrix effects, limit of detection, stability of standards and storage time of sample containing fibre. Finally, three breath samples from a healthy patient were analysed for acetone and isoprene using the 65 μm polydimethylsiloxane/divinylbenzene fibre. The results, 38 ± 1.1 and 10 ± 1.4 $\text{nmol}\,\text{l}^{-1}$ for acetone and isoprene, respectively were within the range found in healthy people. Figure 5.12 shows the total ion chromatogram of a breath sample taken 1 hour after the consumption of approximately 0.5 l of beer and stored for 3 hours on dry ice prior to analysis.

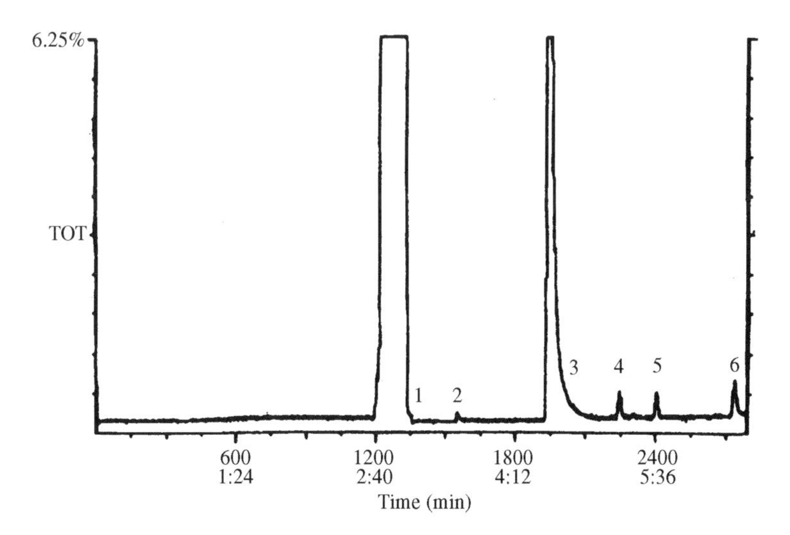

Figure 5.12 GC-MS chromatogram of a breath sample. Breath sample taken after the consumption of alcohol, 65 μm Carbowax/DVB fibre. Peak assignment: 1 = air, CO_2; 2 = acetaldehyde; 3 = ethanol; 4 = acetone; 5 = isopropene; 6 = carbon disulfide. Extraction time, 30 s; desorption time, 15 s; desorption temperature, 200 °C. Reprinted with permission from Grote and Pawliszyn, *Analytical Chemistry*, **69** (1997) 587. Copyright (1997) American Chemical Society

5.5.3 ANALYSIS OF CIGARETTE SMOKE CONDENSATE

SPME coupled to GC-(SIM)MS has been utilised for the quantitative determination of phenolic compounds in cigarette smoke condensate.[24] Sixteen compounds (phenol, *o*-cresol, *m*-cresol, *p*-cresol, 2-methoxyphenol, 2,6-dimethylphenol, 2,4-dimethylphenol, 2,5-dimethylphenol, 3-ethylphenol, 4-ethylphenol, 2,4,6-trimethylphenol, 4-methoxyphenol, 3-methoxyphenol, vanillin, 1-naphthol and 2-naphthol) present in Kentucky 2R1F (from the Tobacco and Health Research Institute) and five commercial cigarette brands were determined. To obtain the smoke condensate, cigarettes were smoked according to the Federal Trade Commission method.[25] Smoke samples were collected on Cambridge pads, treated with 2′-hydroxyacetophenone, and then extracted with 15 ml of HCl–KCl buffer solution using a wrist action shaker for 1 hour. The aqueous solution was decanted and then extracted using an 85 μm polyacrylate fibre for 1 hour with stirring. The results (Figure 5.13) compare favourably with those reported previously for the Kentucky Reference cigarettes and commercial brands.

5.5.4 HEADSPACE SPME OF CINNAMON

The classification of the botanical origin of cinnamon by SPME and GC was reported by Miller *et al.*[26] True cinnamon (in the UK and its former colonies and most of Europe) is defined as the dried inner bark of the species *Cinnamomum zeylanicum* Nees and is imported largely from Sri Lanka, the Seychelles and Madagascar. In the USA and some other countries the term cinnamon is used for a wider range of botanical sources including *Cinnamomum cassia* Blume, *Cinnamomum loureirii* Nees and *Cinnamomum burmanii* Blume, generically referred to as cassia, and originating mainly from Indonesia, China and Burma. True cinnamon has a characteristically different flavour to cassia and is considered to be a superior flavouring agent and hence usually commands a higher price. It is therefore important to be able to differentiate between the different botanical sources because of economic importance. For SPME both non-polar (7 μm and 100 μm polydimethylsiloxane) and polar (85 μm polyacrylate) fibres were used. The sample (10 mg) of cinnamon or cassia was placed into a 10 ml test tube with PTFE-lined silicone septa for headspace sampling. The tube was then immersed in an oil bath at 70 °C for 15 min prior to extraction. The SPME fibre was exposed to the sample for 5 min under the hot conditions; after removal the fibre was inserted into the injector of the GC for analysis (desorption, 270 °C for 2 min). The highest yield of semivolatiles was obtained using the 100 μm polydimethylsiloxane fibre. The results were analysed using principal component factor analysis. It was found that the presence of benzyl benzoate and eugenol in true cinnamon, absent in cassia, and coumarin and *δ-cadinene* in cassia, absent or in trace amounts in true cinnamon, are the most important factors in distinguishing between the two spices.

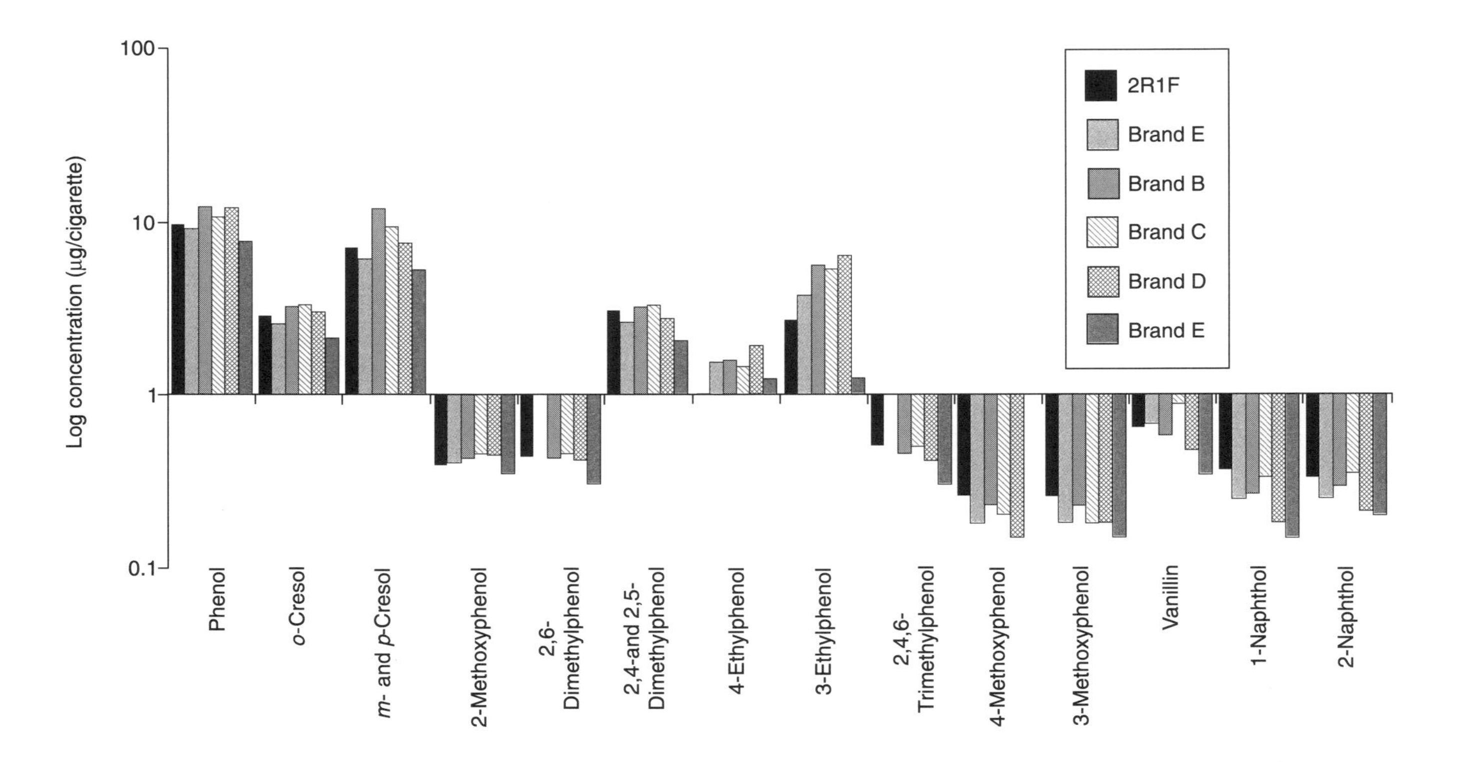

Figure 5.13 SPME of cigarette smoke condensate. From Clark and Bunch, *Journal of Chromatographic Science*, **34** (1996) 272

5.5.5 TETRAETHYLLEAD AND INORGANIC LEAD IN WATER

A method for the determination of tetraethyllead (TEL) and ionic lead in water by SPME followed by GC has been described.[27] The TEL is extracted from the headspace above the sample using SPME while ionic lead requires derivatisation with sodium tetraethylborate to form TEL prior to extraction. The analytical procedure was evaluated with respect to pH (pH 4.5 optimal), amount of derivatising reagent (0.4 ml of a 1% solution of sodium tetraethylborate), stirring conditions (for vials of 17 mm i.d. and 16 mm long stir bars, 1800 rpm was required), and extraction time (15 min) for derivatisation of ionic lead. The detection limit for TEL was 100 ppt when using FID and 5 ppt with ion trap mass spectrometry.

5.5.6 SOLID PHASE MICROEXTRACTION-ELECTRODEPOSITION DEVICE

A modified SPME device in which the 'fibre' is used as the electrodeposition working electrode for the determination of diamines (putrescine and cadaverine) by GC has been reported.[28] Normally, GC methods for the analysis of diamines require a derivatisation step to either facilitate extraction into an organic solvent or improve detector response. Methods using HPLC are also limited in terms of sensitivity because of a lack of a strongly absorbing chromophore in the compounds. In the SPME approach a pencil 'lead' acts as both an adsorbent for SPME and a working electrode for electrodeposition. The working electrode is immersed in a pH 8 borate buffer solution and a −1.70 V potential versus Ag/AgCl applied. This results in an electrochemical reduction of buffer solution protons. As a result of this reduction process, a higher pH at the surface of the electrode occurs compared to the bulk solution. Solution protonated diamines are then converted to their free bases and subsequently retained on the electrode surface. The electrode is then removed from the solution and introduced into the hot injector of the GC for thermal desorption, separation and detection. This approach was applied to the analysis of a Burgundy wine sample, pretreated by passing through a C18 silica column to remove interferences. Using the approach it was possible to determine, using the standard additions method, that the levels of putrescine and cadervine in the wine sample were 270 and 11 ppb, respectively.

5.5.7 ANALYSIS OF POLAR ANALYTES USING DERIVATISATION/SPME

Trace analysis of fatty acids in water and/or air samples was done using SPME with derivatisation of the target analytes.[29] Derivatisation allows the target analytes to be converted to less polar and more volatile species prior to GC analysis. This has the additional benefit of increasing the fibre coating/water or fibre coating/air partitioning coefficient which leads to improved SPME efficiency and method sensitivity. The derivatisation was done in three distinctly separate ways: in the sample matrix; in the SPME fibre coating; and, in the GC injection port. Pentafluorobenzyl bromide and (pentafluorophenyl)diazoethane (PFPDE) were used to derivatise short-chain fatty acids in sample matrices. Diazomethane and pyrenyldiazomethane (PDAM) were used for effective in-fibre derivatisation of long-chain and short-chain fatty acids, respectively. Tetramethylammonium hydroxide and tetramethylammonium hydrogen sulphate were used to convert long-chain fatty acids into their volatile methyl esters via in-injector derivatisation. For solution samples all six reagents were used; for air samples PDAM and PFPDE only were used.

The exposure limit for short-chain fatty acids in industrial air is 25–30 $\mu g\, l^{-1}$ while exposure limits in indoor air are usually 100–1000 times lower (0.025–0.3 $\mu g\, l^{-1}$) therefore it is essential for SPME that sensitivities in this region are obtained. The authors report[29] that the limit of detection for C2–C5 fatty acids in air are within this specification provided derivatisation is used. Figure 5.14 compares the results obtained using two different derivatisation reagents and the limit of detection obtained without derivatisation. The results indicate that either derivatisation reagent can be used to monitor industrial and indoor air quality.

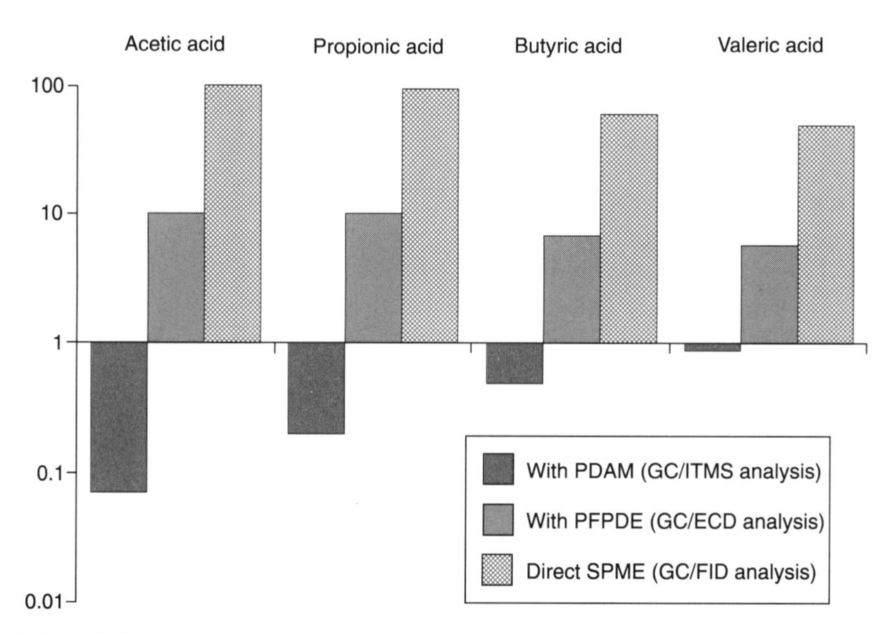

Figure 5.14 Comparison of polyacrylate coated SPME for the analysis of C2–C5 fatty acids in air using derivatisation/SPME with PDAM and PFPDE and direct SPME: limits of detection (ng/l). Reprinted with permission from Pan and Pawliszyn, *Analytical Chemistry*, **69** (1997) 196. Copyright (1997) American Chemical Society

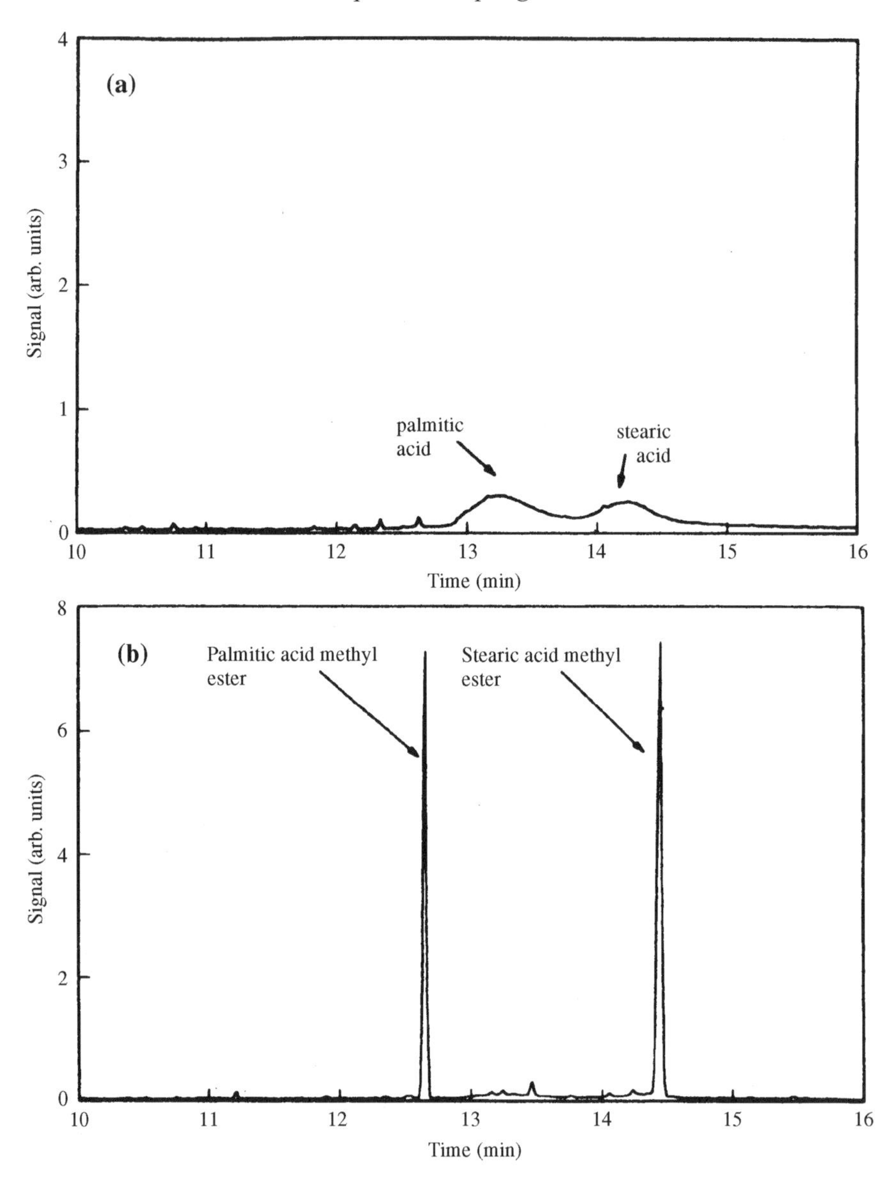

Figure 5.15 GC-FID analysis of (a) underivatised palmitic and stearic acids using direct SPME and (b) methylated palmitic and stearic acids using in-fibre derivatisation. Chromatographic conditions: fibre, PDMS; column, 30 m, 0.25 mm i.d., 1 μm, SPD-5; flow rate 2 ml min^{-1}; oven temperature, 60–280 °C at 20 °C min^{-1}; injector and FID temperature, 300 °C. Note: the *y*-axis in (a) is half of that in (b). Reprinted with permission from Pan and Pawliszyn, *Analytical Chemistry*, **69** (1997) 196. Copyright (1997) American Chemical Society

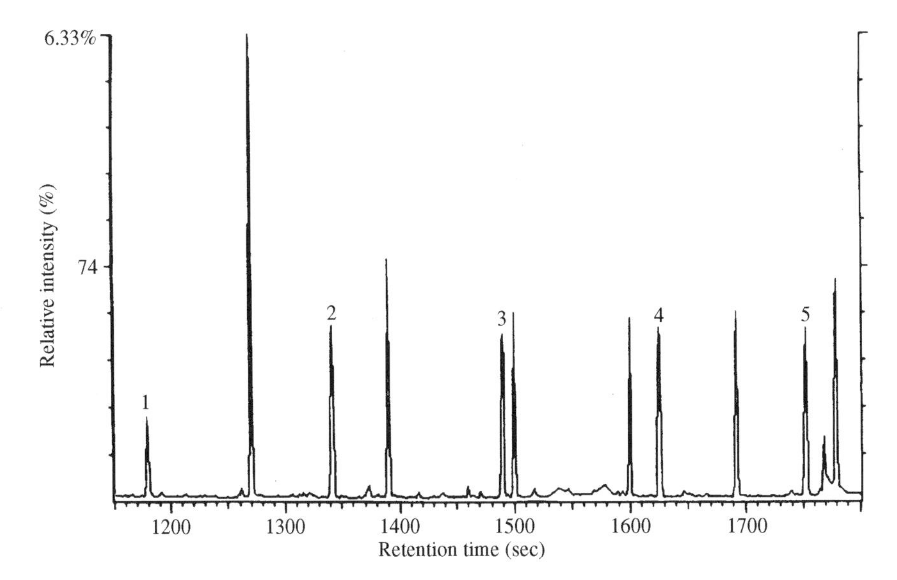

Figure 5.16 Reconstructed selected ion plot with the base peak of *m/z* 74, for the derivatisation of: 1 = C14; 2 = C16; 3 = C18; 4 = C20; 5 = C22 fatty acids with TMA-HSO_4. The actual peaks shown are methyl esters. Chromatographic conditions: fibre, PA; column 30 m, 0.25 mm i.d., 0.25 μm, SPD-5; flow rate 1 ml min^{-1}; oven temperature, 50 °C for 3 min to 270 °C at 7 °C min^{-1}; injector temperature, 300 °C; transfer line, 250 °C; mass range, 45–400 amu. Reprinted by permission from Pan and Pawliszyn *Analytical Chemistry*, **9** (1997) 196. Copyright (1997) American Chemical Society

Limits of detection for short-chain fatty acids using in-matrix derivatisation with either PDAM, PFB-Br or PFPDE were determined. In each case the result is compared to limits of detection obtained without derivatisation. The results indicate that in-matrix derivatisation with PFPDE using the polyacrylate fibre and analysis with GC-ECD gave the lowest limit of detection for C2–C3 acids (0.6–0.8 ng ml^{-1}), whereas PFB-Br gave the lowest limits of detection for C4 and C5 acids (0.1–0.5 ng ml^{-1}). In all cases the derivatisation techniques improved the sensitivities by up to three orders of magnitude.

In-fibre derivatisation with diazomethane was applied to long-chain fatty acids in aqueous solutions. Initially, the polyacrylate fibre is placed in an aqueous sample containing the fatty acids. After sufficient extraction time, the fibre is then withdrawn and inserted into the headspace of another vial containing diazomethane/diethyl ether to carry out the derivatisation. Derivatisation was found to be complete at room temperature within 20 min. Figure 5.15 shows the chromatograms for palmitic and stearic acids: (a) underivatised sample using direct SPME only, and (b) methylated palmitic and stearic acids using in-fibre derivatisation. The dramatic effect of derivatisation is clearly seen.

Preliminary experiments for the in-injector derivatisation of C10–C22 acids with ion-pair reagents TMA-OH and TMA-HSO_4 was reported using GC with ion trap mass spectrometric detection. It was reported that the method was not suitable for the derivatisation of C10 and C12 acids. It was found, however, that the method, using an injector temperature of 300 °C, was suitable for the SPME/derivatisation of C14–C22 acids (Figure 5.16).

REFERENCES

1. R. Eisert and K. Levsen, *J. Chromatogr.*, **733** (1996) 143.
2. D. Louch, S. Motlagh and J. Pawliszyn, *Anal. Chem.*, **64** (1992) 1187.
3. D.W. Potter and J. Pawliszyn, *J. Chromatogr.*, **625** (1992) 247.
4. C.L. Arthur, L.M. Killam, S. Motlagh, M. Lim, D.W. Potter and J. Pawliszyn, *Environ. Sci. Technol.*, **26** (1992) 979.
5. I. Valor, C. Cortada and J.C. Molto, *J. High Res. Chromatogr.*, **19** (1996) 472.
6. Z. Zhang and J. Pawliszyn, *Anal. Chem.*, **65** (1993) 1843.
7. B. MacGillivray, J. Pawliszyn, P. Fowlie and C. Sagara, *J. Chromatogr. Sci.*, **32** (1994) 317.
8. I.J. Barnabas, J.R. Dean, I.A. Fowlis and S.P. Owen, *J. Chromatogr.*, **705** (1995) 305.
9. A.A. Boyd-Bowland and J. Pawliszyn, *J. Chromatogr.*, **704** (1995) 163.
10. R. Young, V. Lopez-Avila and W.F. Beckert, *J. High Res. Chromatogr.*, **19** (1996) 247.
11. A.A. Boyd-Bowland, S. Magdic and J.B. Pawliszyn, *Analyst*, **121** (1996) 929.
12. T. Gorecki, R. Mindrup and J. Pawliszyn, *Analyst*, **121** (1996) 1381.
13. K.D. Buchholz and J. Pawliszyn, *Anal. Chem.*, **66** (1994) 160.
14. J.R. Dean, W.R. Tomlinson, V. Makovskaya, R. Cumming, M. Hetheridge and M. Comber, *Anal. Chem.*, **68** (1996) 130.
15. K.J. Hageman, L. Mazeas, C.B. Grabanski, D.J. Miller and S.B. Hawthorne, *Anal. Chem.*, **68** (1996) 3892.
16. H. Daimon and J. Pawliszyn, *Anal. Comm.*, **33** (1996) 421.
17. K.J. James and M.A. Stack, *J. High Resolut. Chromatogr.*, **19** (1996) 515.
18. J. Chen and J. Pawliszyn, *Anal. Chem.*, **67** (1995) 2530.
19. A.A. Boyd-Bowland and J. Pawliszyn, *Anal. Chem.*, **68** (1996) 1521.
20. D. De la Calle Garcia, S. Magnaghi, M. Reichenbacher and K. Danzer, *J. High Res. Chromatogr.*, **19** (1996) 257.
21. L.K. Ng, M. Hupe, J. Harnois and D. Moccia, *J. Sci. Food Agric.*, **70** (1996) 380.
22. C. Grote and J. Pawliszyn, *Anal. Chem.*, **69** (1997) 587.
23. A. Manolis, *Clin. Chem.*, **29** (1983) 5.
24. T.J. Clark and J.E. Bunch, *J. Chromatogr. Sci.*, **34** (1996) 272.
25. H.C. Pillsbury, C.C. Bright, K.J. O'Conner and F.W. Irish, *J. Assoc. Off. Anal. Chem.*, **52** (1969) 458.
26. K.G. Miller, C.F. Poole and T.M.P. Pawlowski, *Chromatographia*, **42** (1996) 639.
27. T. Gorecki and J. Pawliszyn, *Anal. Chem.*, **68** (1996) 3008.
28. E.D. Conte and D.W. Miller, *J. High Res. Chromatogr.*, **19** (1996) 294.
29. L. Pan and J. Pawliszyn, *Anal. Chem.*, **69** (1997) 196.

BIBLIOGRAPHY

J. Pawliszyn, *Solid Phase Microextraction*, John Wiley & Sons Ltd, New York (1997).

II

SOLID SAMPLES

6

Solid Sample Preparation

The extraction of organic pollutants from solid matrices, i.e. soils, sludges, etc. is important for the maintenance of effective environmental safety. The potential safety implications of contaminated land are related to (a) its future usage and (b) input into natural (ground and surface) water supplies. The former is of particular concern if the land site was contaminated because of previous industrial usage which may have resulted in accidental or deliberate spillage of toxic materials. Both these cases will have led to environmental pollution of the land. In an ideal world, no contamination of the land would occur, however, this is not the case. Perhaps more realistically, is the uptake by industrial manufacturers of so-called 'green' policies which implies that they may modify their production of a product to reduce its impact on the environment. In addition, the treatment of waste prior to disposal into rivers, streams and other water courses is now governed by legislation in developed countries.

If environmental land contamination can be proven, then it is the responsibility of the land-owner (this may in itself require legal action to identify who is responsible) to effect action. This so-called remediation may take the form of treatment, containment or removal of the pollutants from the soil all of which require the monitoring of the soil prior to and after treatment.[1] These approaches will now be considered.

Containment

A common approach to contaminated land remediation is to remove the polluted material from its source and relocate it to a landfill site. This involves placing the waste in a lined pit. The purpose of the lining is to prevent loss of toxic material. However, over a long time period this is unlikely and the release of toxic materials to ground water should be considered. A properly controlled site should have the capability to monitor leachate and also landfill gas emissions, e.g. CH_4.

Treatment

Incineration is probably the most common approach for treatment of organic pollutants. In this situation, the waste is passed into an incinerator at high temperature (~ 1200 °C) with an adequate oxygen supply to aid combustion of the waste. Unfortunately, this type of treatment can lead to a secondary level of pollution (e.g. acid rain) arising from emissions from the incinerator. Strict regulation of these emissions has led to improvements in the technology required to clean-up flue gas emissions. In addition, the products of the incineration, i.e. ash, are contaminated with inorganic-type residues which also require disposal. If the incinerator is not working efficiently there is the possibility that highly toxic organic pollutants (i.e. dioxins) can be synthesised and then released into the environment. However, an efficient incinerator is ably suited for the disposal of organic pollutants with good efficiency. As a side issue, the heat generated from the combustion of the waste can be used to produce electricity.

An alternative strategy is to compost the waste after shredding and separation of the putrifiable material, mixing it with other material, e.g. sewage sludge, and then relying on microbial action to degrade the waste into a suitable compost. The compost can then be used to grow plants as an alternative to peat. As far as contaminated soil is concerned this is not the preferred option as it does not necessarily have any influence on the level of pollution present (see later).

New/alternative approaches to treatment of contaminated soil are also becoming available. These new approaches include: plasma arc destruction, which utilises high temperatures (up to 45 000 °C) to destroy the most inert of organic pollutants; and, superheated water, which uses water heated to 370 °C at 220 bar pressure and in the presence of oxygen can oxidise organic pollutants to CO_2 and H_2O. The benefits of these newer technologies versus their financial costs and environmental impact is still emerging.

At the present time in the UK, the government is committed to the 'suitable for use' approach to contaminated sites.[2] This implies that remediation is only necessary when contamination poses unacceptable risks to the environment (i.e. ground water and surface water quality) and health (i.e. people, animals and plants).[3] These risks, at least in the UK, are based on trigger contaminant concentrations, established through the Department of the Environment, by the Interdepartmental Committee for the Redevelopment of Contaminated Land (ICRCL).[4] For the most commonly found environmental organic pollutants the trigger concentrations relate to the final use of the contaminated land. An example is given in Table 6.1.[3,4]

This implies that when the land is to be used for domestic usage, e.g. gardens and allotments for growing vegetables, the greatest protection to the end-user is given with the lowest levels of pollutants only allowed. This is extremely important as the vegetables grown on the land may have the capability to accumulate the pollutants and hence provide a potential health risk to the consumer (directly if a

Table 6.1 Trigger contaminant concentrations[3,4]

Contaminant	Proposed use of land	Trigger concentration/mg kg^{-1} air-dried soil	
		Threshold	Action
Polycyclic aromatic hydrocarbons	Domestic gardens, allotments, play areas	50	500
Polycyclic aromatic hydrocarbons	Landscaped areas, buildings, hard cover	1000	10000
Phenols	Domestic gardens, allotments	5	200
Phenols	Landscaped areas, buildings, hard cover	5	1000

human or indirectly if vegetables are given to a farm animal in the first instance). In the case where the land is to be used as part of the construction industry, i.e. office blocks, car parks, the permitted levels are higher and hence the soil is more contaminated. This is perhaps understandable insomuch as the contaminated land provides no immediate health risk to people. There is, of course, still a potential environmental risk if the pollutants leach into ground water. (Note: It should be noted, however, that the values quoted are continually under review and are provided by ICRCL as guidelines only and are used only to demonstrate the complexity of the situation; other countries obviously have their own guidelines.)

6.1 INTRODUCTION

The removal of organic pollutants from contaminated soil sites for the purposes of assessing the environmental impact of the pollutants is required to assess what level of remediation is required. Assuming that the appropriate sampling and storage of samples has occurred (Chapter 1) the analyst is required to select appropriate techniques to desorb the organic pollutants from the soil. This in itself is not straightforward bearing in mind the wide choice of techniques that are currently available. This wide choice of analytical techniques available for separation of pollutants from solid environmental matrices and their selection will be discussed in detail in subsequent chapters. However, it may be necessary to be aware of other considerations, e.g. how will you know that total recovery of the pollutant has occurred? What influence does the soil matrix have on the retention of the pollutants? and, which techniques have approved methods? Each of these areas will now be considered.

How will you know that total recovery of the pollutant has occurred?

This is not as easy to answer as you might at first think. The essence of the problem is that if you are considering the level of pollution on a contaminated land site, on which the pollutant has been released some time ago, it will have undoubtedly have undergone some form of interaction (chemical, physical) or degradation. This may mean that (a) it is no longer in its original chemical form or (b) has undergone some form of immobilisation. The former may not be a cause of concern, if the chemical or microbial alteration has produced a less harmful substance, which is of no environmental interest, i.e. is not on the 'red' list of toxic chemicals.

The analytical extraction process can therefore be influenced by the nature of the soil-pollutant interactions. In order to address the problem the analyst has various avenues of approach. The first is that a particular sample can be extracted using two different methods or techniques and the results compared. This as you can guess is time-consuming and may not be conclusive, particularly for a new technique which may require additional sample work-up procedures. A solution to this approach is to participate in certification or proficiency testing schemes, e.g. CONTEST operated by the Laboratory of the Government Chemist (LGC) in the UK. In these cases, each laboratory selects their most appropriate extraction method, sample work-up procedure (if required) and method of quantitation. The results generated are then sent to the coordinator who compares all results, as supplied, and produces an overall report that allows each individual laboratory to assess their performance. The outcome of this approach may be the production of a certified reference material (CRM). This will occur if the original organisers of the certification scheme is an appropriate body, e.g. National Institute of Science and Technology (NIST) in the USA, the Community Bureau of Reference (or BCR) in Brussels or the LGC. If this is the case, then other non-participating laboratories can purchase CRMs for which certifiable data is provided. By selecting appropriate CRMs, e.g. soil contaminated with phenols, any new methods of extraction and/or quantitation can be evaluated and results compared with the certificate. This does not entirely answer our original question, however. It is probable that the methods of extraction used in the certification scheme will be the most common methods available, e.g. Soxhlet extraction for pollutants from soils, and hence the same as those available in your own laboratory. So, it is possible that exhaustive extraction has not occurred. A problem with the use of CRMs is their limited availability for a wide range of sample types and analytes. Ongoing research and development and certification schemes do ensure, however, that new CRMs are produced at fairly frequent intervals.

It is not uncommon to find in the scientific literature, due in some respects to the limited availability of CRMs and their cost, that extraction techniques are unfairly compared. Therefore, it is surprisingly common to find that new extraction techniques are compared using sample matrices spiked with the pollutant of interest. Two approaches to spiking exist. The first, 'spot' spiking is where an aliquot of analyte in organic solvent is added to a relatively large mass of soil. The small volume of solvent removed by evaporation either by heat or ambient temperature (depending on the solvent) and the sample extracted and analysed. The reader will not be surprised to find that the results quoted by the preferred extraction technique are excellent and hence compare extremely favourably with existing extraction methodology. The alternative spiking approach is 'slurry' spiking. In this case, an aliquot of analyte is added in a relatively large volume of organic solvent to the soil matrix. The solvent is then stirred, typically overnight in a fume cupboard, to remove solvent and provide a (hopefully) homogeneously spiked sample. While this is a better approach to 'spot' spiking, in that it does allow the analyte some time to interact with the soil, it still does not represent the 'real' environmental situation.

What influence does the soil matrix have on the retention of the pollutants?

To understand the question requires knowledge of the chemistry and mineralogy of inorganic and organic soil components to comprehend the diversity of chemical reactions that pollutants may undergo in the soil matrix. The type of chemical reactions that may be important are: equilibrium and kinetic processes such as dissolution, precipitation, polymerisation, adsorption/desorption and oxidation-reduction. It is then important to understand how they may affect the solubility, mobility and toxicity of contaminants in soil. While such discussion is extensively beyond the scope of this book it is essential to give a perspective on the diversity of components that we tend to regard as soil.

Soils are complex assemblies of solids, liquids and gases that vary according to location and prevailing weather conditions (particularly important for the liquid and gas components). In general, soil can be divided into inorganic and organic components. The inorganic components are largely made up of minerals. A mineral is described as a natural inorganic component with a definite physical, chemical and crystalline property. Two classes of mineral are important: primary and secondary. Primary minerals are characterised by the fact that they have not been chemically altered since their deposition and crystallisation from molten lava; whereas a secondary mineral has undergone some type of chemical alteration. Both primary and secondary minerals range in particle diameter from clay-sized colloids ($< 2\,\mu m$) to gravel ($> 2\,mm$) and rocks. Commonly found primary minerals are: quartz (SiO_2), and aluminosilicates (feldspars) whereas secondary minerals include clay minerals, i.e. kaolinite, montmorillonite and vermiculite.

In addition to the inorganic components of soil is the soil organic matter (SOM), sometimes called the humus. However, while it is possible to characterise and identify mineral components of soil, the structure and chemistry of SOM is far from definitive. The SOM content ranges from 0.5 to 5% in the surface horizons of mineral soils to 100% in organic soils. In addition, SOM has a high specific surface (up to 800–900 $m^2 g^{-1}$) and a cation exchange capacity (CEC) between 150 and 200 cmol kg^{-1}.[5] Due to this high specific surface and CEC, the SOM is the major sorbent of organic pollutants (also heavy metals and plant nutrients) in soil. The SOM can be subdivided into nonhumic and humic substances. The nonhumic substances have recognisable physical and chemical properties such as carbohydrates, proteins, fats and waxes. As these components are attacked by soil microrganisms they persist in the soil only briefly. Humic substances are relatively high molecular weight substances, coloured in appearance and characterised by their solubility characteristics. The humic substances can be further subdivided into humic acid, fulvic acid and humin. Humic acid is dark in coloration and can be extracted from soil by various reagents and is itself insoluble in dilute acid; fulvic acid is the coloured fraction that remains in solution after removal of humic acid by acidification; and humin is the alkali insoluble fraction of SOM.

To predict the fate and mobility of pollutants in contaminated soils requires information on the desorption of the determinands. For example if the determinand is strongly bound to the soil and minimal desorption occurs or is slow, then the polluted soil offers little in the way of potential risk to the ground water supply. In contrast if the pollutant is easily desorbed it could become mobile and contaminate water supplies. Otherwise, depending on the use of the soil, the persistence of the pollutant on the soil may lead to long-term contamination issues if the soil is used for domestic uses, e.g. growing of vegetables. These issues as well as being of environmental importance have implications for extraction in the analytical laboratory. If a pollutant is likely to be easily desorbed from the soil via ground or surface water, the implication is that it can be easily extracted from the soil by a suitable solvent. In contrast the persistence of pollutants on soil also might lead to less than quantitative recovery in the laboratory.

Many soil chemical processes are time-dependent.[5] It can therefore be inferred that the desorption of pollutants from soils in the extraction process are time-dependent, i.e. kinetically controlled.

Which extraction techniques have approved methods?

The use of any extraction technique requires some external justification that the technique is capable of fulfilling its function, i.e. extraction of pollutants from soil. External justification for the use of extraction techniques is achieved through regulatory agencies, e.g. United States Environmental Protection Agency (USEPA)

Table 6.2 Current USEPA methods for extraction of pollutants from solid environmental matrices

USEPA method	Extraction technique	Applicable to	Additional comments
3540	Soxhlet extraction	Semivolatiles and nonvolatile organics from soils, relatively dry sludges and solid wastes	Considered a rugged extraction method because it has very few variables that can adversely affect extraction recovery
3541	Automated Soxhlet extraction	Polychlorinated biphenyls, organochlorine pesticides and semivolatiles from soils, relatively dry sludges and solid wastes	Allows equivalent extraction efficiency to Soxhlet in 2 hours
3545	Accelerated solvent extraction	Semivolatiles and nonvolatile organics from soils, relatively dry sludges and solid wastes	Extraction under pressure using small volumes of organic solvent
3550	Ultrasonic extraction	Semivolatiles and nonvolatile organics from soils, relatively dry sludges and solid wastes	Considered to be less efficient than other methods detailed; rapid method that uses relatively large solvent volumes. Not appropriate for use with organophosphorus compounds as it may cause destruction of the target analytes during extraction
3560	Supercritical fluid extraction	Semivolatile petroleum hydrocarbons from soils, sludges and waste	Normally uses pressurised CO_2 with additional small volumes of organic solvents; limited applicability; relatively rapid extractions
3561	Supercritical fluid extraction	Polynuclear aromatic hydrocarbons from soils, sludges and waste	As method 3560

and their equivalents in other countries, who provide the necessary information accumulated (usually) over many years that approves a particular technique. The USEPA is regarded as one of the main sources of approved methods for extraction, and hence will be detailed here. The USEPA approved extraction methods for organic pollutants from soils are shown in Table 6.2. The current absence of an approved method for the extraction of pollutants from soils by microwave-assisted extraction is noted.

REFERENCES

1. B.J. Alloway and D.C. Ayres, *Chemical Principles of Environmental Pollution*, Blackie Academic and Professional, London (1993).
2. *Framework for Contaminated Land*, London, Department of the Environment, November (1994).
3. C. Grundy, *Chemistry in Britain*, **33**(2) (1997) 33.
4. *Guidance on the assessment and redevelopment of contaminated land*, 2nd edition, ICRCL guidance note 59/83, London, Department of the Environment (1987).
5. D.L. Sparks, *Environmental Soil Chemistry*, Academic Press, Inc., New York (1995).

7

Liquid–Solid Extraction

The extraction of pollutants from soils is most commonly done using traditional liquid–solid extraction methods. The most common of these approaches is Soxhlet extraction (and modifications thereof) with shake-flask extraction offering an even cheaper alternative while sonication provides a rapid approach to extraction.

7.1 INTRODUCTION

The utilisation of liquid–solid extraction for the removal of analytes from environmental matrices is the oldest of the methods discussed. Soxhlet extraction was introduced by Baron von Soxhlet in the mid-nineteenth century.

The methods of liquid–solid extraction can be conveniently divided into those for which heat is required (Soxhlet/Soxtec) and those methods for which no heat is added, but which utilise some form of agitation i.e. shaking or sonication. The use of sonication can lead to a gentle form of heating generated by the use of the sonic probe or bath. In Soxhlet extraction only one sample can be extracted per set of apparatus, but it is possible to operate with as many sets of apparatus as space in a fume cupboard allows. Multiple extractions can be easily carried out using the shake-flask approach using simple laboratory shakers. With sonication, the speed of extraction (approximately 3 min per extraction; 3 extractions carried out on the same sample) allows multiple extractions to be carried out rapidly.

7.2 EXPERIMENTAL

For Soxhlet extraction two variations of the apparatus are possible (Figure 7.1). The difference between the two designs is whether the solvent vapour is allowed to cool by passing to the outside of the apparatus or whether it remains within the body of

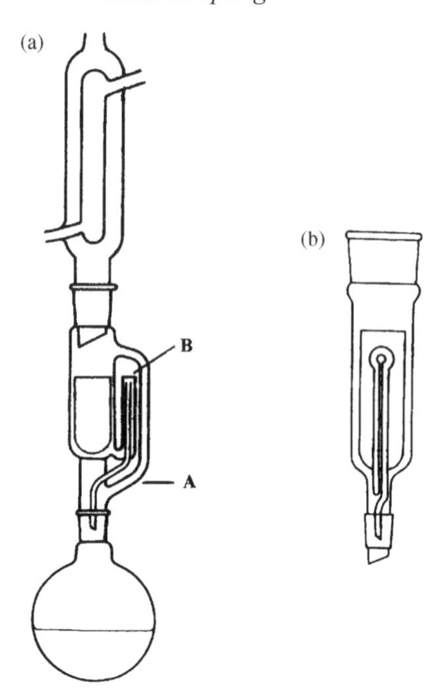

Figure 7.1 Soxhlet apparatus: (a) solvent vapour passes external to the sample-containing thimble, this results in cooled organic solvent passing through the sample, the extraction process is relatively slow; (b) solvent vapour passes surrounds the sample-containing thimble, hot organic solvent allows more rapid extraction

the apparatus. The mode of operation of both types is the same and only the former will be described in detail.

The solid sample (approx. 10 g if a soil) and a similar mass of anhydrous sodium sulphate are placed in the porous thimble (cellulose) which in turn is located in the inner tube of the Soxhlet apparatus. The apparatus is then fitted to a round-bottomed flask of appropriate volume containing the organic solvent of choice, and to a reflux condenser. The solvent is then boiled gently using an isomantle; the solvent vapour passes up through the tube marked (A), is condensed by the reflux condenser, and the condensed solvent falls into the thimble and slowly fills the body of the Soxhlet apparatus. When the solvent reaches the top of the tube (B), it syphons over into the round-bottomed flask the organic solvent containing the analyte extracted from the sample in the thimble. The whole process is repeated frequently until the pre-set extraction time is reached. As the extracted analyte will normally have a higher boiling point than the solvent it is preferentially retained in the flask and fresh solvent recirculates. This ensures that only fresh solvent is used to extract the analyte from the sample in the thimble. A disadvantage of this approach is that the organic solvent is below its boiling point when it passes through the sample contained in the thimble. In practice this is not necessarily a problem as Soxhlet extraction is normally done over long time periods, i.e. 6, 12, 18 or 24 hours.

Automated Soxhlet extraction or Soxtec utilises a three-stage process to obtain rapid extractions. In stage 1, a thimble containing the sample is immersed in the boiling solvent for approximately 60 min. After this (stage 2) the thimble is elevated above the boiling solvent and the sample extracted as in the Soxhlet extraction approach. This is done for up to 60 min. The final stage (stage 3) involves the evaporation of the solvent directly in the Soxtec apparatus (10–15 min). This approach has several advantages including speed (it is faster than Soxhlet, approximately 2 hours per sample), the fact that it uses only 20% of the solvent volume of Soxhlet extraction and the sample can be concentrated directly in the Soxtec apparatus.

Conventional liquid–solid extraction, in the form of shake-flask extraction is carried out by placing a soil sample into a suitable glass container, adding a suitable organic solvent, and then agitating or shaking (rocking or circular action) for a prespecified time period. After extraction, the solvent-containing analyte needs to be separated from the matrix by means of centrifugation and/or filtration. In some instances, it may be advisable to repeat the process several times with fresh solvent and then combine all extracts.

Sonication involves the use of sound waves to agitate the sample immersed in organic solvent. The preferred approach is to use a sonic probe although a sonic bath can be used. The sample is placed in a suitable glass container and enough organic solvent added to cover the sample. The sample is then sonicated for a short time period, typically 3 min, using the sonic bath or probe. After extraction, the solvent containing the analyte is separated by centrifugation and/or filtration and fresh solvent added. The whole process is repeated three times and all the solvent extracts combined.

In all cases, some form of solvent-containing analyte reduction of volume is required. This can take the form of liquid–liquid extraction or solid-phase extraction. These approaches, as well as reducing the solvent volume and hence preconcentrating the analyte, may also include inherent clean-up processes.

7.3 SELECTED METHODS OF ANALYSIS

As the methods described in this chapter are often used as the benchmark techniques it is not particularly common to find papers that focus only on their usage. However, it is extremely common to find that papers that are proposing alternative extraction methods/techniques to compare results use liquid–solid extraction, e.g. Soxhlet. Accordingly it was decided not to focus on particular environmental pollutants in this chapter but to highlight selected applications of the various liquid–solid extraction approaches. No attempt is made to provide a comprehensive literature survey.

7.3.1 SOXHLET

An investigation into the extraction and clean-up of sewage sludge-amended soils prior to chromatographic analysis for polychlorinated biphenyls (PCBs) was described by Folch *et al.* (1996).[1] Three different Soxhlet solvent systems (200 ml) were evaluated: hexane, dichloromethane and hexane-acetone (41 : 59, v/v). Each soil (1 g sewage sludge or 10 g sludge-amended soil) was Soxhlet extracted for 6 h at a rate of 4–6 cycles h^{-1}. The results (Figure 7.2) indicate that either solvent system is suitable. However, as hexane is non-polar and dichloromethane is halogenated both these solvents were rejected in favour of hexane–acetone (41 : 59, v/v). The addition of recently precipitated sulphur in the bottom of the Soxhlet thimbles was advantageous for the removal of interfering sulphur. Clean-up of samples was done using a microcolumn of florisil.

7.3.2 SOXTEC (AUTOMATED SOXHLET)

An evaluation of the Soxtec procedure for extracting organic compounds from soils and sediments was reported by Lopez-Avila *et al.* (1993).[2] This study was carried

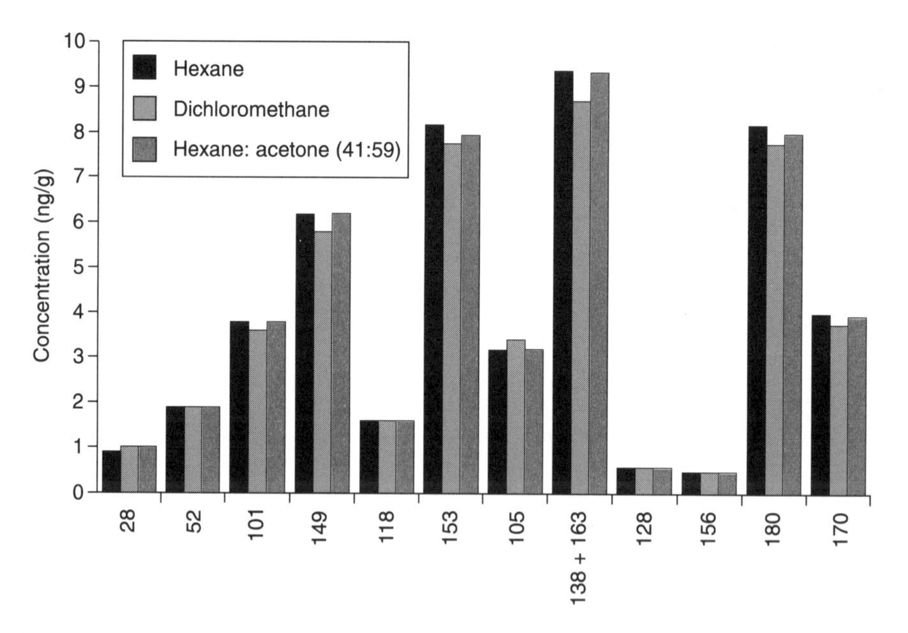

Figure 7.2 PCB recoveries from 15% sewage sludge-amended soil using different extraction solvents. From Folch *et al.*, *Journal of Chromatography*, **719** (1996) 121

out on 29 target compounds (seven nitroaromatic compounds, three haloethers, seven chlorinated hydrocarbons and twelve organochlorine pesticides) from spiked sandy clay loam and clay loam. The study involved two solvent systems (hexane-acetone and dichloromethane-acetone, both as $1+1$ v/v). In addition, the effect on analyte recovery from five factors was considered i.e. matrix type, spike level, anhydrous sodium sulphate addition, total extraction time and immersion/extraction time ratio. Of the five factors investigated, matrix type, spike level and total extraction time had a significant effect on method performance at the 95% confidence interval for 16 out of 29 target compounds (four compounds were not recovered at all). Anhydrous sodium sulphate and immersion/extraction time ratio had no statistical significance at the 95% confidence interval, except for *trans*-chlordane. The data indicate that the two solvent systems are equally suitable for Soxtec extraction. In addition, 64 basic/neutral/acidic compounds were spiked onto clay loam at the 6 mg kg^{-1} level. The recoveries obtained for these compounds were as follows: 20 had recoveries of >75%, 22 between 50 and 74%, 12 between 25 and 49% and 10 had recoveries of <25%. Three certified reference materials were also analysed for PAHs, i.e. a soil sample, SRS103-100 (Fisher Scientific) and two marine sediments (HS-3 and HS-4) from NRCC, using hexane–acetone $(1+1)$. The results are shown in Figures 7.3–7.5, respectively. The

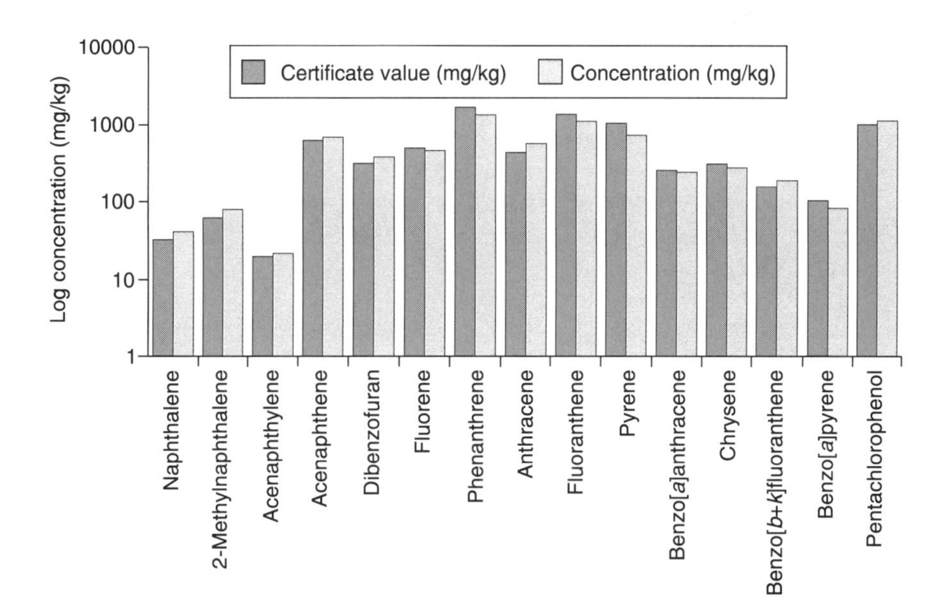

Figure 7.3 Soxtec extraction of compounds from SRS103-100 standard reference soil using hexane–acetone $(1+1)$ and the following conditions: immersion time, 45 min; extraction time, 45 min; sample size, 10 g. From Lopez-Avila *et al.*, *Journal of the Association of Official Agricultural Chemists*, **76** (1993) 864

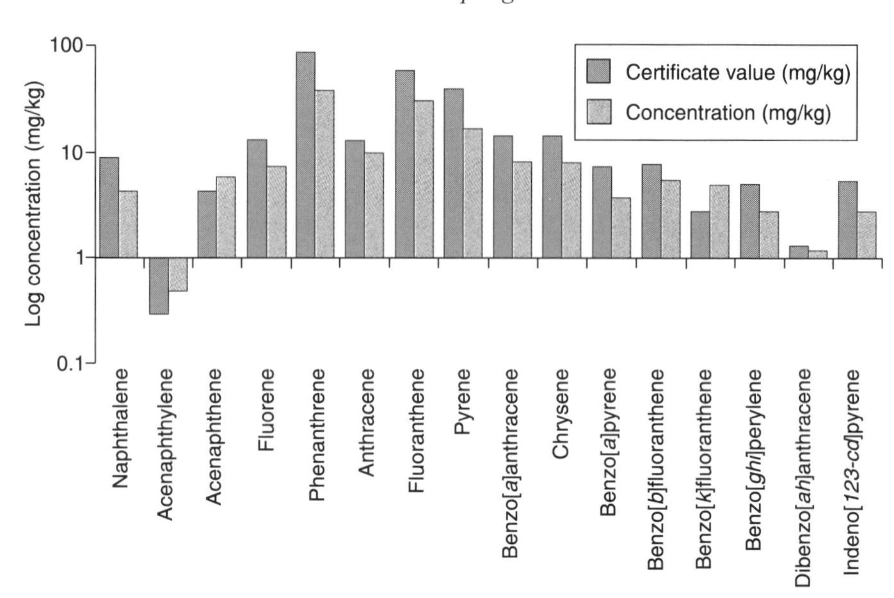

Figure 7.4 Soxtec extraction of PAHs from HS-3, marine sediment reference material using hexane–acetone (1 + 1) and the following conditions: immersion time, 45 min; extraction time, 45 min; sample size, 10 g. From Lopez-Avila *et al.*, *Journal of the Association of Official Agricultural Chemists*, **76** (1993) 864

recoveries ranged from 69–131% for the SRS103-100 reference soil sample, from 43–175% for the HS-3 marine sediment, and 59–133% for HS-4 marine sediment. The authors conclude by proposing Soxtec as an alternative extraction technique to Soxhlet and sonication procedures currently (paper published in 1993) recommended by USEPA.

7.3.3 SHAKE FLASK

In 1990, Cheng[3] evaluated the sequential extraction of two soils for radio-labelled atrazine using mixed solvents. The soil was treated with the ^{14}C-labelled atrazine and incubated under controlled temperature and pressure for three months. The soils were then exhaustively extracted (either separately or sequentially with each solvent) using 0.05 M $CaCl_2$, 1 M KCl, acetone, acetonitrile, methanol, acetone + $CaCl_2$, and methanol + $CaCl_2$ until no significant amount of atrazine ($< 1\%$) could be removed with a solvent. For each extraction a soil:solvent ratio of 1 : 10 was used. Extraction was done by shaking the soil–solvent mixture for 30 min. This was followed by centrifugation, decanting of the solvent and finally evaporation of the solvent. From the results (Table 7.1) it is evident that a particular solvent system will consistently remove a similar portion of atrazine. Once removed, further

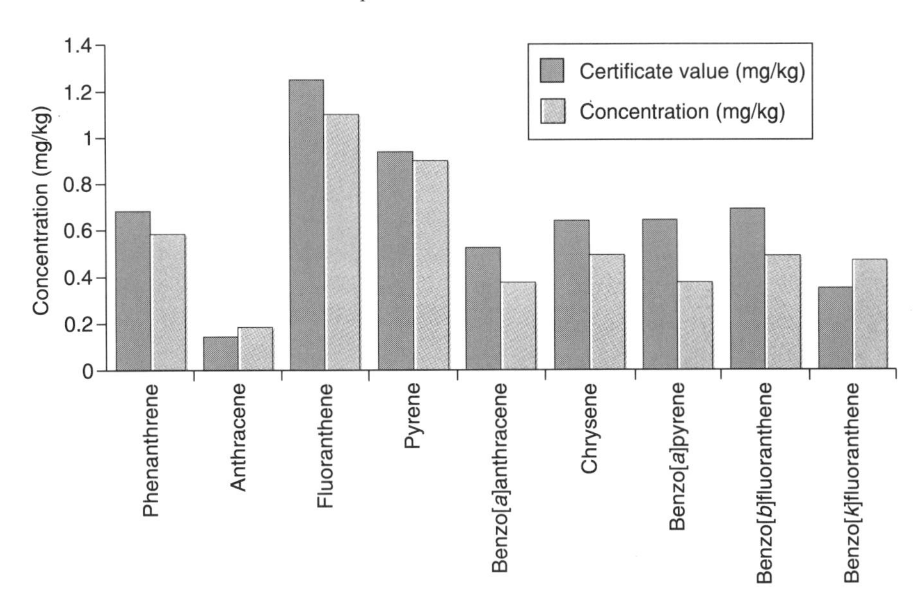

Figure 7.5 Soxtec extraction of PAHs from HS-4, marine sediment reference material using hexane–acetone (1 + 1) and the following conditions: immersion time, 45 min; extraction time, 45 min; sample size, 10 g. From Lopez-Avila *et al.*, *Journal of the Association of Official Agricultural Chemists*, **76** (1993) 864

treatment with the same solvent will not remove any more atrazine. This indicates the specificity of the solvent for bond breaking. The most effective solvent combination was a salt solution with a miscible organic solvent. In each case the combined extraction solvent was more effective than the sum of the individual solvent ($CaCl_2$ or methanol/acetone). It was postulated, based on previously reported work[4,5] that hydrogen bonding between the chemical and soil surface can be reduced when organic solvents are present in an aqueous solution. This would indicate that the dominant mechanism for atrazine retention on this soil was hydrogen bonding. These results were compared with those obtained using sequential extraction which involved the use of all the solvent systems studied, with the exception of 1 M KCl. It is noted (Table 7.1) that the sequential approach was only marginally more effective than the use of a mixed solvent of acetone or methanol and $CaCl_2$. It was suggested by the author that the remaining atrazine not recovered from the soil by the solvents used was incorporated into the soil humus.

The effect of extraction systems on the recovery of clomazone, a soil-applied herbicide, from aged soil samples was reported by Kirksey and Mueller (1995).[6] Seven solvent extraction systems were used. Each involved the shaking of moist aged soil (40 g) with either acetonitrile or methanol (80 ml solvent volume) for different times (1 or 16 h) with or without repeat extractions. The soil used was a

Table 7.1 Extractability[a] of ^{14}C-residue from ^{14}C-atrazine treated soils[3]

Solvent	Number of determinations	Soil E1	Soil E2	Soil N3	Soil N4
0.05 M $CaCl_2$	8	6.6	7.4	9.7	9.8
1 M KCl	2	8.0	8.9	11.2	11.0
Acetone	8	13.0	18.7	19.3	22.6
Acetonitrile	8	13.1	17.7	20.3	22.0
Methanol	8	26.2	32.2	37.6	41.4
Acetone + $CaCl_2$	8	50.8	59.4	57.7	66.0
Methanol + $CaCl_2$	5	52.6	58.8	59.7	61.2
Sequential[b]	29	60	66	63	67

[a] Samples were exhaustively extracted by one solvent before extraction by a subsequent solvent. Data are averages of four replicates.
[b] Sequentially with acetone, methanol and acetonitrile in different orders, followed by acetone + $CaCl_2$ and methanol + $CaCl_2$.
Soil E1/2: Eschweiler soil is a sandy loam-degraded loess; it has a pH of 5.2 and contains 1.35% C, 0.112% N, 12.0% clay, 28.4% silt and 59.6% sand.
Soil N3/4: Neuhofen Neu soil is a sandy soil; it has a pH of 7.0 and contains 3.02% C, 0.255% N, 8.3% clay, 7.6% silt and 84.1% sand.

silt loam with a pH of 6.0 and an organic matter content of 2.0%. The field sample was taken 14 days after application of the herbicide. No attempt was made to air dry the soil. It was noted that after the application of the herbicide to the soil that 3 cm of rainfall had occurred. The results, based on triplicate samples, are shown in Figure 7.6. The highest recoveries were obtained using the following extraction systems: 16 h acetonitrile followed by 1 h acetonitrile (458 ± 18 ng g^{-1} and 16 h methanol followed by 1 h methanol (448 ± 14 ng g^{-1}). All other extraction systems gave incomplete clomazone recovery. Acetonitrile was preferred because of chromatographic compatibility and ease of rotary evaporation.

7.3.4 SHAKE-FLASK VERSUS SONICATION

Smith (1978)[4] compared a variety of solvent systems (six) for the extraction of four herbicides (benzoylprop-ethyl, nitrofen, profluralin and tri-allate) from several field soils that has been weathered for at least 6 months at different locations in Saskatchewan, Canada. The composition and physical characteristics of each soil studied was determined (Table 7.2). Air-dried soil samples (20 g) were extracted ($n = 2$) with 50 ml of the appropriate solvent and either shaken for 1 h on a wrist-action shaker or sonicated for 2 min. The solvents used and their proportions (v/v) were: acetone, methanol, acetonitrile + water (9 + 1), acetonitrile + water (7 + 3), acetonitrile + water + glacial acetic acid (8 + 1.8 + 0.2) and acetonitrile + water + glacial acetic acid (7 + 2.7 + 0.3). The results (Figures 7.7 and 7.8) indicate

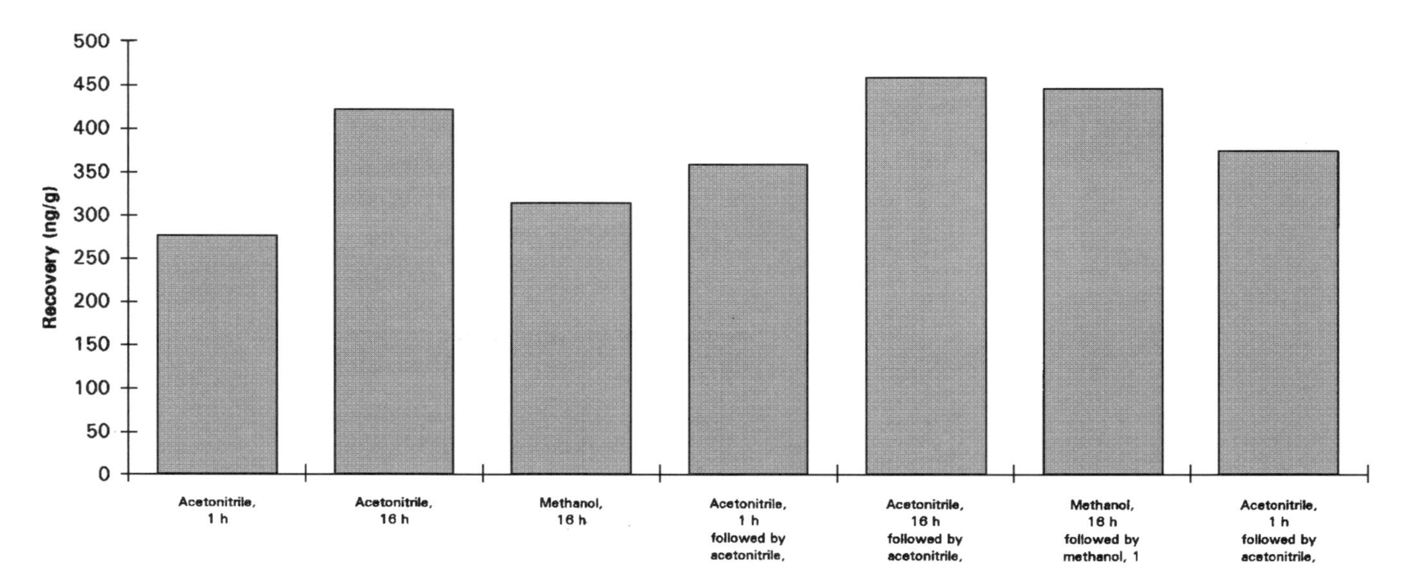

Figure 7.6 Recovery of clomazone from 0 to 8 cm soil zone at 14 days after treatment at a rate of 1.12 kg clomazone ha^{-1}. From Kirksey and Mueller, *Journal of the Association of Official Agricultural Chemists*, **78** (1995) 1519

Table 7.2 Soil properties[4]

Soil	pH	Sand (%)	Silt (%)	Clay (%)	Organic content (%)	Field capacity (% moisture)
Jameson, sandy loam	7.5	85	9	6	3.2	11
Regina, heavy clay	7.7	5	26	69	4.2	40
Melfort, silty clay	5.2	32	38	30	11.7	36

that either acetonitrile + water (7 + 3) or aqueous acetonitrile containing glacial acetic acid recovered the highest concentrations of the herbicides from the aged soils. Except for tri-allate it was apparent that the presence of acetic acid did not significantly enhance the extraction efficiencies. It was postulated that this lack of effect reflects the absence of ionic binding between the herbicides and the soils. The author referred to other work which was in agreement with these findings. It was suggested that the use of an acetonitrile + water + acetic acid (7.0 + 2.7 + 0.3)

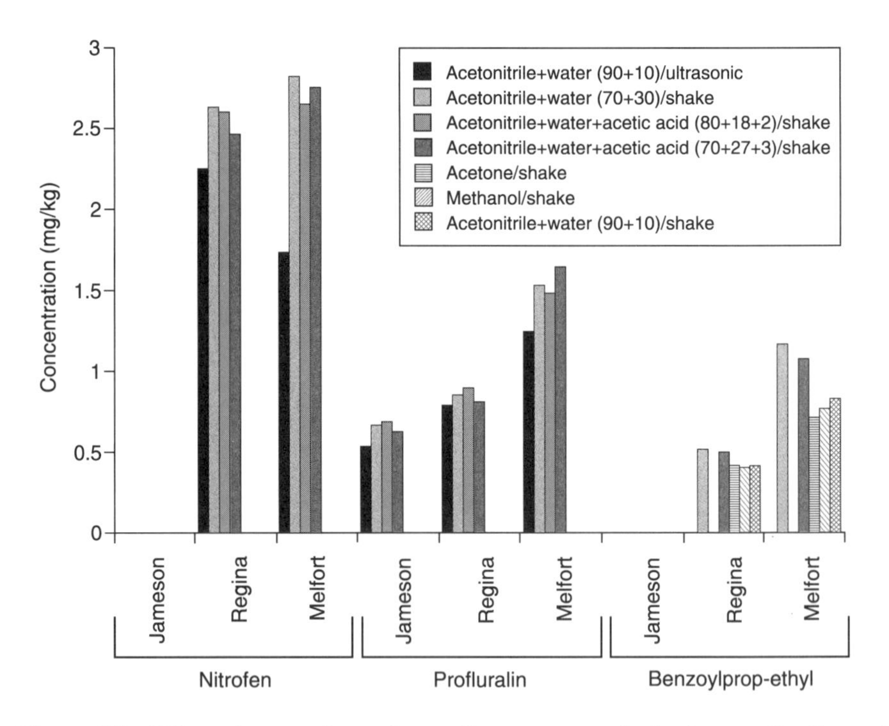

Figure 7.7 Effect of extraction solvents for recovery of nitrofen, profluralin and benzoylprop-ethyl after 6 months on weathered soils. From Smith, *Pesticide Science*, **9** (1978) 7

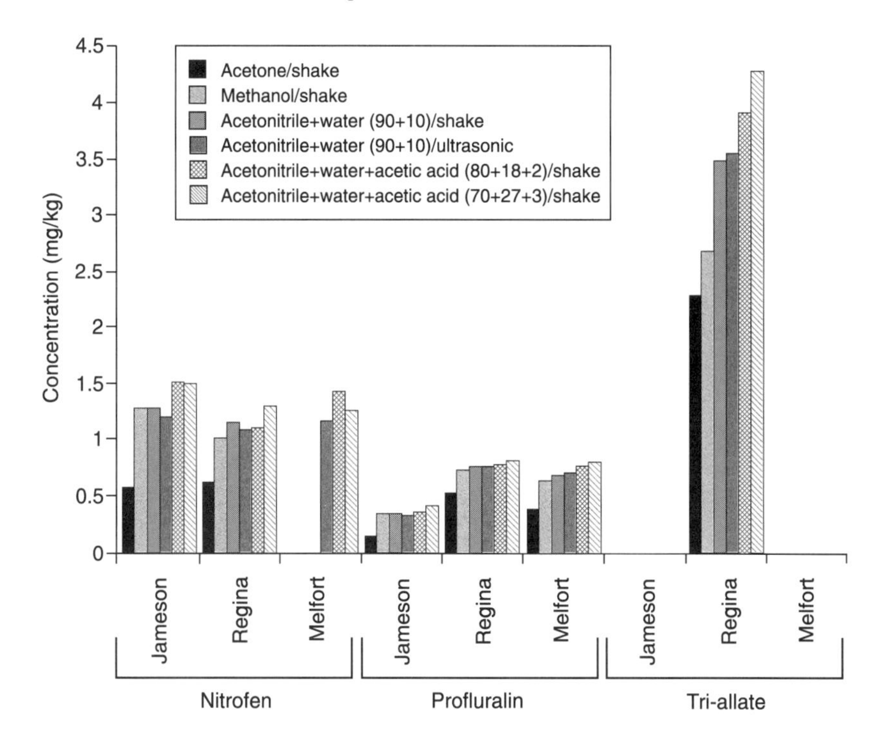

Figure 7.8 Effect of extraction solvents for recovery of nitrofen, profluralin and tri-allate after 17 months on weathered soils. From Smith, *Pesticide Science*, **9** (1978) 7

solvent mixture was suitable for the extraction of both acidic herbicides residues as well as non-acidic herbicide residues from weathered Saskatchewan field soils. The presence of water was considered to be beneficial for good extraction recoveries.

7.3.5 SHAKE-FLASK VERSUS REFLUX (SOXHLET)

In 1980 Cotterill[5] compared the efficiency of methanol with acetonitrile and chloroform to extract herbicide residues from soil. The author reported that acetone was not included in this study as it usually produces extracts containing high levels of extraneous soil constituents. Soils were obtained from nine locations in the UK, eight of which contained weathered residues (fluometuron, lenacil, linuron, metribuzin, propyzamide and simazine). In addition, the eight soils containing pesticide residues were characterised (Table 7.3). The soil samples were taken three months after application. All soils were air-dried and passed through a 3-mm sieve, then stored at $-15\,^{\circ}$C until ready for extraction, when they were thawed using a microwave oven.

Table 7.3 Soil properties[5]

Soil	pH	Sand (%)	Silt (%)	Clay (%)	Organic carbon (%)	Field capacity (% moisture)
1	5.9	12	32	56	25	38.9
2	6.2	71	15	14	1.3	16.1
3	5.9	74	15	11	19	61.9
4	6.8	77	18	5	50	76.3
5	7.1	74	16	10	1.5	16.7
6	5.9	72	13	15	2.1	21.6
7	5.8	78	17	5	56	78.4
8	7.0	73	11	16	1.6	16.6
9[a]	—	—	—	—	—	—

[a] Insufficient soil, a sandy loam, was available for analysis.

Soil samples (25 g) were extracted with the appropriate solvent (total volume 50 ml in each case) either by refluxing for 8 h or by shaking on a wrist-shaker for 1 h. The solvents used were acetonitrile + water (9 + 1, v/v) or chloroform for reflux conditions, and acetonitrile + water (9 + 1, v/v), methanol or methanol + water (4 + 1, v/v) for cold shaking. Extracts were either filtered or decanted off (chloroform extracts only) after settling. The results (Figure 7.9) indicate that the highest recoveries ($n = 3$), irrespective of soil type, were obtained in all cases after shaking with aqueous methanol as solvent. The effect of water on the extraction of weathered simazine residues in soil 2 was evaluated. It was found that the optimum recovery (0.78 mg kg^{-1}) was only obtained using a methanol + water ratio of 4 + 1 (v/v). The author postulated that the presence of water helps to break down the structure of the soil, allowing the extractant to work on a greater surface area. To investigate this, further samples of soil 2 were air-dried and sieved through mesh sizes of 150, 300, 1700 and 3000 μm before extraction with methanol only. It was found that as the particle size decreased, and hence the surface area increased, so the recovery of simazine increased. The maximum extracted was 0.75 mg kg^{-1}, using methanol only and a soil particle size of 150 μm. This recovery was considered not to be significantly different to that obtained using a methanol + water ratio of 4 + 1 (v/v) as the extraction solvent (mesh size 3000 μm).

7.3.6 SONICATION VERSUS SOXHLET

The extraction of polychlorinated biphenyls (PCBs) from soils has been investigated with particular emphasis on a comparison of the extraction systems (technique, solvent and clean-up procedures) used in an inter-laboratory study.[7] The samples were prepared using slurry spiking and 25 g subsamples distributed to

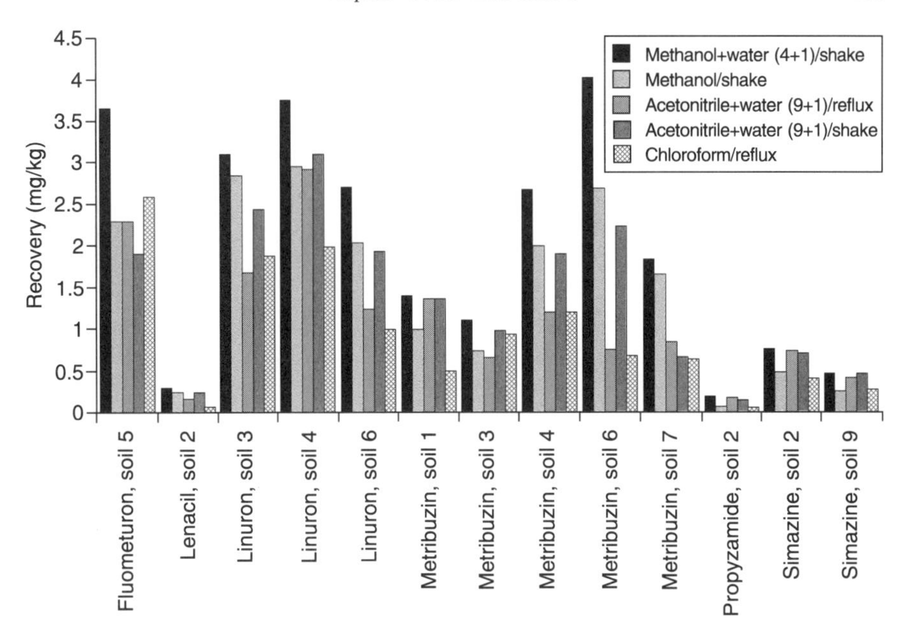

Figure 7.9 Solvent extraction of herbicide residues from weathered soils. From Cotterill, *Pesticide Science*, **11** (1980) 23

129 laboratories. Two standard methods of extraction were evaluated; EPA method 3540 for Soxhlet extraction and EPA method 3550 for sonication extraction. The solvents recommended for extraction by both methods are dichloromethane, hexane–acetone (1 + 1) and toluene–methanol (9 + 1). This last solvent system was not used by any of the laboratories. However, other solvent systems were evaluated including dichloromethane–acetone (1 + 1), dichloromethane–hexane (1 + 1), hexane and isooctane. The results indicated that Soxhlet extraction is more reliable and gives more accurate results than sonication. The combination of Soxhlet extraction with chlorinated or polar solvents is the most efficient and robust method. However, this method also produces greater amounts of co-extractives from the soil which require extensive clean-up prior to PCB analysis. The use of non-polar organic solvents, such as hexane, is not recommended for the extraction of PCBs from a polar matrix, such as soil.

7.3.7 OTHER APPROACHES

A redesigned simultaneous steam distillation-solvent extraction (SDE) apparatus was proposed by Seidel and Lindner[8] as a universal sample enrichment technique

Table 7.4 Comparison between SDE, SFE and Soxhlet extraction for the determination of hexachlorobenzene (HCB) in soil[8]

	Steam distillation-solvent extraction	Supercritical fluid extraction	Soxhlet extraction
Process time	60 min	25 min	4 h
Further clean-up of soil extracts	not required	not required	solid phase extraction/Florisil
Clean-up time (min)	—	—	45
Concentration time (min)	30	5	20
Overall analysis time (h)	1.5	0.5–1.3	6
Analysis time/person (h) (in parallel mode)	0.5 (3 SDE devices)	0.5–1.3 (1 SFE chamber)	1 (8 Soxhlets)
HCB recovery (%)[a]	100	98	55
Standard deviation (%)	2.7	3.5	7.5
Detection limit (ppb)	0.05	0.1	0.5
Maximum sample amount (g)	100	5	100

[a] Level of HCB 10 ppb (SDE reported as 100% recovery).

for organochlorine pesticides from environmental (and biological) samples. Although steam distillation is one of the oldest known enrichment techniques for volatile compounds it is only rarely used for OCPs. The technique is also known as on-line, continuous, or concurrent SDE.[9,10] The technique is based on extraction using non-polar organic solvents which are lighter than water, e.g. petroleum ether (40–60 °C). In this paper the redesigned SDE apparatus (details given) was compared with existing concurrent distillation-extraction apparatus, Soxhlet extraction and SFE. Soxhlet extraction was done on a 10 g sample of soil, mixed with 10 g of anhydrous sodium sulphate and extracted with 180 ml of petroleum ether (40–60 °C) for 4 h. For SFE a 5 g soil sample was mixed with 2.5 g of anhydrous sodium sulphate and methanol organic modifier added (350 μl). The extraction was done at 40 °C and 380 bar in static mode for 15 min followed by dynamic mode for a further 5 min. For the SDE a 60 min extraction time was used together with 50 or 100 g of sample (experimentally verified). Each technique was compared (Table 7.4) and the recovery of hexachlorobenzene from a naturally contaminated agricultural soil reported. One of the main advantages of SDE is the high purity of the obtained organic extract, which eliminates further clean-up prior to GC-ECD analysis. This modified approach was suggested as a versatile and promising method for monitoring anthropogenic contaminants in a range of matrices.

REFERENCES

1. I. Folch, M.T. Vaquero, L. Comellas and F. Broto-Puig, *J. Chromatogr.*, **719** (1996) 121.
2. V. Lopez-Avila, K. Bauer, J. Milanes and W.F. Beckert, *J. AOAC Int.*, **76** (1993) 864.
3. H.H. Cheng, *Int. J. Environ. Anal. Chem.*, **39** (1990) 165.
4. A.E. Smith, *Pestic. Sci.*, **9** (1978) 7.
5. E.G. Cotterill, *Pestic. Sci.*, **11** (1980) 23.
6. K.B. Kirksey and T.C. Mueller, *J. AOAC Int.*, **78** (1995) 1519.
7. D.E. Kimbrough, R. Chin and J. Wakakuwa, *Analyst*, **119** (1994) 1283.
8. V. Seidel and W. Lindner, *Anal. Chem.*, **65** (1993) 3677.
9. R.A. Flath and R.R. Forrey, *J. Agric. Food Chem.*, **25** (1977) 103.
10. S.T. Lickens and G.B. Nickerson, *Proc. Am. Soc. Brew. Chem.* (1964) 5.

8

Supercritical Fluid Extraction

Supercritical fluid extraction (SFE) relies on the diversity of properties exhibited by the supercritical fluid to (selectively) extract analytes from solid, semi-solid or liquid matrices. The important properties offered by a supercritical fluid for extraction are: (a) good solvating power; (b) high diffusivity and low viscosity, and (c) minimal surface tension.

SFE can be operated in two modes: off-line and on-line. While a significant part of the early work was done on utilising the compatibility of SFE to chromatographic separation, the lack of robust instruments and the inflexibility of such systems has probably precluded their continued usage. The situation is exactly the opposite for off-line SFE. In this situation, the flexibility of off-line operation allows the analyst to focus on the sample preparation only. This allows SFE to be optimised to maximise analyte recovery, to process larger samples without fear of overloading the chromatographic column, a certain amount of freedom of choice pertaining to the analysis, e.g. GC, HPLC, IR, etc., and finally, that the analytical measurement instrument is available to analyse other samples.

Historical perspective

A supercritical fluid is any substance above its critical temperature and pressure. The discovery of the supercritical phase is attributed to Baron Cagniard de la Tour in 1822.[1] He observed that the boundary between a gas and a liquid disappeared for certain substances when the temperature was increased in a sealed glass container. Subsequent work by Hannay and Hogarth in 1879[2–4] demonstrated the solvating power of supercritical ethanol (T_c 243 °C; P_c 63 atm) by studying the solubilities of cobalt(II) chloride, iron(III) chloride, potassium bromide and potassium iodide. They noticed that the solubilities of the chlorides were much higher than would be expected by their vapour pressure. In addition, they also noted that increasing pressure caused the metal chlorides to dissolve while decreasing the pressure caused precipitation. Despite this early promise, however, the utility of supercritical fluids for extraction was left dormant for many years. The pioneering work of Francis[5] on liquified gases is noteworthy. In this work Francis compiled an extensive list of solubilities for 261 compounds in near-critical carbon dioxide. A major development in the use of supercritical carbon dioxide was the filing of various patents by Zosel between 1964 and 1976 on the decaffeination of coffee.[6] This development has led to the growth of applications in the engineering field of supercritical fluid technology. A major development was the installation by Kraft General Foods of a decaffeination plant for their Maxwell House Coffee Division in 1978, which uses an extraction cell with a height of 25 m. This utilisation of supercritical fluids for pilot and full-scale plant operations has continued to develop and diversify. In these situations supercritical fluids are being used for a wide diversity of applications (Table 8.1). It is thus not surprising to find that supercritical fluids encompass a multidisciplinary field that includes engineers, chemists, food scientists, material scientists and workers in biotechnology, agriculture and environmental control.

Development in the use of supercritical fluids for analytical usage is historically routed in chromatography where the pioneering work of Novotny and Lee introduced capillary SFC in 1981.[8] Analytical scale SFE first appeared commercially in the mid-1980s.

Table 8.1 Industrial uses of SFE[7]

Year	Operator	Materials processed
1982	SKW/Trotsberg	Hops
1984	Fuji Flavor Co.	Tobacco
	Barth and Co.	Hops
	Natural Care Byproducts	Hops, Red pepper
1986	CEA	Aromas, pharmaceuticals
1987	Messer Griesheim	Various
1988	Nippon	Tobacco
	Takeda	Acetone residues from antibiotics
	CAL-Pfizer	Aromas
1989	Clean Harbors	Waste waters
	Ensco, Inc.	Solid wastes
1990	Jacobs Suchard	Coffee
	Raps and Co.	Spices
	Pitt-Des Moines	Hops
1991	Texaco	Refinery wastes
1993	Agrisana	Pharmaceuticals from botanicals
	Bioland	Bone
	US Air Force	Aircraft gyroscopic components
1994	AT&T	Fibre optics rods

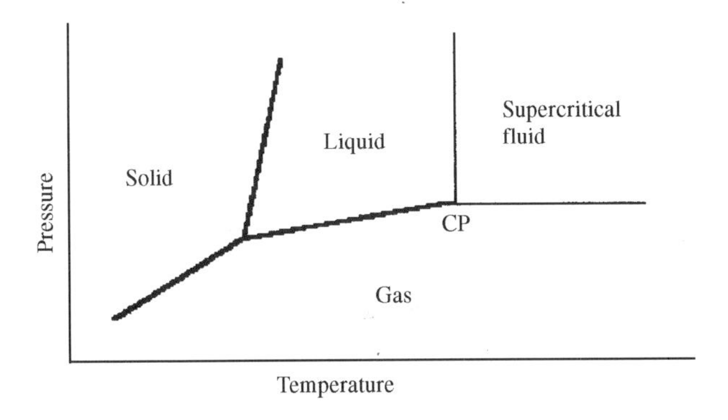

Figure 8.1 Phase diagram for a pure substance

8.1 DEFINITION OF A SUPERCRITICAL FLUID

A phase diagram for a pure substance is shown in Figure 8.1. In this diagram can be seen the regions where the substance occurs, as a consequence of temperature or pressure, as a single phase, i.e. solid, liquid or gas. The divisions between these regions are bounded by curves indicating the co-existence of two phases, e.g. solid–gas corresponding to sublimation; solid–liquid corresponding to melting; and finally, liquid–gas corresponding to vaporisation. The three curves intersect at the triple point where the three phases co-exist in equilibrium. At the critical point, designated by both a critical temperature and a critical pressure, no liquefaction will take place on raising the pressure and no gas will be formed on increasing the temperature. It is this defined region, which is by definition, the supercritical region. Table 8.2 lists the critical temperature and pressures for a range of substances which are suitable for SFE.

After consideration of the Table 8.2, the choice of supercritical fluid is often determined on the basis of practical issues, such as availability with high purity and

Table 8.2 Critical properties of selected substances

Substance	Critical temperature (°C)	Critical pressure (atm)	Critical pressure (psi)
Xenon	16.7	59.2	847.0
Carbon dioxide	31.1	74.8	1070.4
Ethane	32.4	49.5	707.8
Nitrous oxide	36.6	73.4	1050.1
Chlorodifluoromethane	96.3	50.3	720.8
Ammonia	132.4	115.0	1646.2
Methanol	240.1	82.0	1173.4
Water	374.4	224.1	3208.2

low cost. The low critical temperature and moderate critical pressure of carbon dioxide make it the substance of choice for most analytical applications of SFE. However, the non-polar nature of carbon dioxide (it has no permanent dipole moment) means that for a significant portion of applications the solvent strength of the supercritical fluid is inadequate. This problem is circumvented by the addition of a polar organic solvent or modifier to the supercritical fluid. Various approaches are possible and include the addition of organic solvent directly to the sample in the extraction cell, the use of pre-mixed cylinders (a commercially available cylinder of carbon dioxide to which has been added organic solvent) or the use of a second pump so that carbon dioxide and organic solvent can be mixed in-line prior to the extraction vessel. The latter is the preferred choice.

8.2 INSTRUMENTATION FOR SUPERCRITICAL FLUID EXTRACTION

The basic components of an SFE system (Figure 8.2) are: a supply of high purity carbon dioxide; a supply of high purity organic modifier; two pumps; an oven for the extraction vessel; a pressure outlet or restrictor; and a suitable collection vessel for quantitative recovery of extracted analytes. Commercial instruments are available from a range of manufacturers or agents.

As noted previously, carbon dioxide is frequently the supercritical fluid of choice, in spite of its lack of a permanent dipole moment. Therefore, it is not uncommon in practice to have to resort to the addition of an organic modifier to increase the polarity of the solvent mixture. Probably the most common modifier

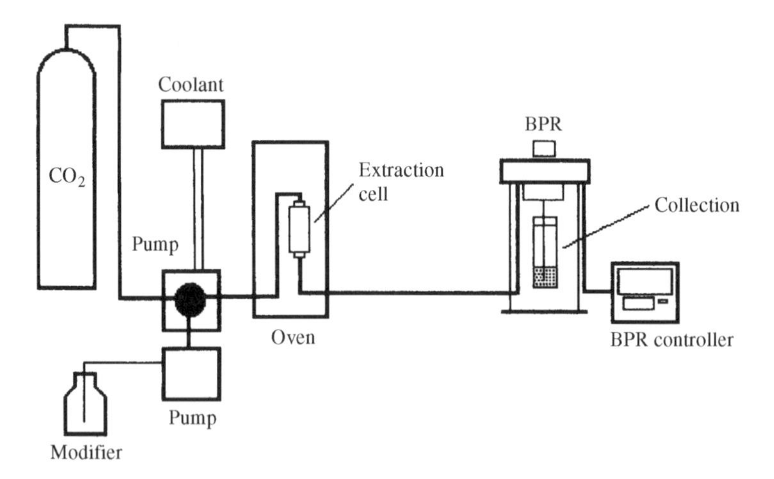

Figure 8.2 Schematic diagram of the basic components of a supercritical fluid extraction system

used is methanol. However, some workers have investigated the influence of modifiers on analyte recoveries. In addition, modifiers have been used with particular purposes in mind other than solvation and polarity, e.g. reactive modifiers have been used to perform in-situ derivatisation. The use of carbon dioxide and a modifier may result in the user having a solvent mixture which is below the critical point. However, while this is technically a definition problem, the user does not usually worry, provided quantitative recoveries are obtained.

Carbon dioxide can be obtained in a range of purities (Table 8.3).[9] It is important that the level of impurities encountered in the carbon dioxide do not interfere with the subsequent analysis. Continued development by the gas suppliers has resulted in carbon dioxide being available which does not contain flame ionisable detectables, i.e. hydrocarbons and electron capture detectables, e.g. halogenated material. The carbon dioxide is supplied in a cylinder fitted with a dip tube. A typical cylinder will comprise liquid carbon dioxide at the bottom of a vertical positioned cylinder and vapour at the top. The presence of the dip tube allows liquified carbon dioxide only to be accessible. It is possible to purchase cylinders that contain both carbon dioxide and an organic modifier. However, previous work has shown that the carbon dioxide to modifier ratio varies over the lifetime of operation of the cylinder.[10,11]

The pressurised gas in the cylinder is pumped through the SFE system using either a reciprocating or syringe pump. The ideal features required of the pump are that it can deliver a constant flow rate, $ml\,min^{-1}$, at a suitable pressure, 3500–10 000 psi. Probably the most common pump for SFE, due to its lower cost, is the reciprocating or piston pump. In this situation, the pump head needs to be cooled to maintain liquefaction of the carbon dioxide. This cooling is achieved using either an ethylene-glycol mixture which is pumped using a recirculating bath, low cost cryogenic-grade carbon dioxide or a peltier cooler. In each case, cooling is required to prevent cavitation, i.e. gas entrapment, in the pump head. Modifier addition is most effectively achieved using a second pump which does not require any pump head cooling. The carbon dioxide and modifier are then subsequently mixed using a T-piece. The combination of two pumps allows a high degree of control in terms of modifier addition and flexibility of choice.

In order to establish the critical temperature for the solvent system requires external heating. This is done via an oven in which is located the sample cell or vessel. The ideal temperature range of the oven is up to 100 °C (although as will be shown later temperatures of 200–250 °C may be advantageous). The sample vessel, which has typically been made of stainless steel, must be capable of withstanding high pressures (up to 10 000 psi) safely. It is important to note that the newly prepared and inserted sample-containing extraction vessel will require some time to achieve the temperature of the pre-set oven prior to commencing the extraction. Ideally, the extraction vessel should be capable of insertion into the system without the need for wrenches. This allows ease of use and rapid change-over of samples and does not lead to excessive wear and tear on the pressure fittings. Commercial

Table 8.3 Purity of commercial carbon dioxide[9]

Typical impurities analysis	Industrial grade without dip tube	Industrial grade with dip tube	High purity with dip tube	SFC grade	SFE-SFE grade
Carbon dioxide	> 99.5%	> 99.99%	> 99.997%	99.995%	> 99.999%
Nitrogen	< 500 ppm	< 50 ppm	< 10 ppm	10 ppm	—
Oxygen	< 200 ppm	< 20 ppm	< 10 ppm	7 ppm	—
Water	< 10 ppm	< 10 ppm	< 5 ppm	3 ppm	< 200 ppb
Carbon monoxide	< 1 ppm[a]	< 1 ppm[a]	< 1 ppm[a]	1 ppm[a]	—
Total hydrocarbons[b]	< 10 ppm	< 10 ppm	< 5 ppm	3 ppm	< 100 ppb
Total sulfur[c]	< 0.1 ppm	< 0.1 ppm	< 0.1 ppm	0.1 ppm	—
Total halocarbons[d]					< 0.1 ppb

[a] Limits of detectability.
[b] Expressed as methane.
[c] Expressed as sulphur dioxide.
[d] Relative to aldrin electron-capture detector response.

systems and specialist suppliers can provide suitable extraction vessels. Also, the ideal shape of an extraction vessel is such that it is easier to place the sample in a short squat vessel rather than a long thin vessel. Almost all commercial extraction vessels are a flow-through design that allows fresh, clean supercritical fluid to pass over the sample.

The maintenance of pressure within the extraction vessel is done using either fixed or variable (mechanical or electronically controlled) restrictors. Historically, this has been the weak link in the analytical usage of SFE due to its unpredictability when extracting real samples. Fixed restrictors are typified by the use of narrow fused-silica or metal capillary tubing while variable restrictors by back-pressure regulators. While the use of so-called fixed restrictors based on fused-silica or metal-capillary tubing is the cheapest option their lack of robustness in all but the hands of the SFE enthusiast does preclude their usage, particularly for real samples. For real samples therefore it is more appropriate to use a variable restrictor. A variable restrictor allows a constant, operator-selected flow rate whose preselected pressure is maintained by the size of the variable orifice. So, while variable restrictors are more expensive than fixed restrictors the freedom from problems, such as, blockages is invaluable.

An interesting paper published by King *et al.*[12] investigated whether (a) methods developed on non-commercial apparatus could be transferred to commercial systems, (b) SFE-derived results were instrument dependent, and (c) to determine precision levels for specific extractions on various instrument modules. Four commercial systems were optimised (Lee SFE-703, Hewlett-Packard 7680A, Isco SFE 2-10 and a Suprex Prepmaster) for the extraction of incurred pesticide residues in poultry fat with *in-situ* alumina clean-up, exhaustive delipidation of soybean flakes and pesticides in wheat samples. The results suggested that the selected SFE methods could be successfully translated onto commercial instruments. All of the instruments gave excellent pesticide recoveries, and the reproducibilities obtained were acceptable. Extractions of oils and fats are best done on instruments that allow pressures up to 700 atm to be achieved. The authors finally concluded that some SFE instruments were more amenable to automation than others.

8.3 METHODS OF ANALYSIS: EXTRACTION FROM SOLID SAMPLES

The purpose of this section is not to provide a thorough literature survey of the work on SFE due to its extensive nature. It should be noted that as well as the bibliography at the end of the chapter, numerous reviews on SFE have been published with an environmental bias.[13–18] However, the intention is to select a limited number of applications that highlight certain aspects of the application of SFE to environmental issues. Using this as a criterion, various classes of compounds of environmental importance have been selected and these are polycyclic aromatic hydrocarbons, polychlorinated biphenyls, pesticides and phenols. Finally a section on extraction from aqueous media using SFE is included.

A particular problem with work reported in the literature is that extraction techniques are frequently evaluated against samples that have been freshly spiked with the target analyte(s). In the work reported in this section only data obtained from unadulterated samples and/or certified reference materials is included. However, because of the limited number of papers that address this issue it was considered appropriate to include papers where an attempt to age or investigate matrix properties has been considered.

8.3.1 POLYCYCLIC AROMATIC HYDROCARBONS

Polycyclic aromatic hydrocarbons (PAHs) have been commonly extracted from environmental matrices to demonstrate the feasibility of SFE. It is therefore not uncommon to find a review devoted to the SFE of PAHs.[19] In this paper various aspects of SFE were considered, namely, the use of alternatives to supercritical carbon dioxide, the use of alternative organic modifiers to methanol for improved extraction efficiency, methods to prevent restrictor plugging, collection of analytes after supercritical fluid depressurisation, optimisation of operating conditions and spiking versus native PAH extraction. The paper concluded by recommending an approach for the quantitative SFE of PAHs from environmental matrices. The main recommendations were:

- Use an SFE system with two pumps.
- Choose the most appropriate modifier, combined modifier mixture or reactive modifier.
- Smaller particle size is important for increasing recovery.
- Extraction from native contaminated samples is better for method development.
- Pack the extraction cell appropriately, i.e. use drying agents and copper (for soils with high sulfur content).
- Use a variable restrictor, not a fixed linear restrictor.
- Choose an appropriate collection system, i.e. liquid–solid trap and/or collection solvent.
- Maintain collection solvent at 5 °C.
- Extraction time is an important variable.
- Extraction temperature (up to 250 °C) is an important variable.
- Consider the use of experimental design/chemometric methods for method development.

Literature examples will now be used to demonstrate the applicability of some of these recommendations.

The effects of temperature and pressure on the extraction efficiency of PAHs from a certified reference material (urban air particulate, SRM CRM 1649) and a contaminated soil (USEPA certified, lot no. AQ103) using carbon dioxide only were investigated.[20] At a constant temperature of 50 °C no improvement in extraction efficiencies was noted on either sample when the pressure was altered from 350–650 atm. For the urban air particulate CRM the highest recoveries were obtained when both high temperature (200 °C) and pressure (at least 350 atm) were used. It is, however, important to note that the higher molecular weight PAHs were not efficiently removed even under these extreme conditions. Figure 8.3 demonstrates the improvement in recoveries obtained for the extraction of PAHs from the urban air particulate CRM.

This work was further reported by Hawthorne and Miller (1994)[21] who directly compared Soxhlet extraction with low- and high-temperature supercritical carbon dioxide extraction for the removal of PAHs from railroad bed soil and diesel soot. The Soxhlet extraction (18 h) was done using either DCM for the rail road soil (10 g samples) or DCM for the diesel soot (2 g samples) as the extraction solvent. The soil was mixed 1 : 1 with sodium sulphate. For the SFE sample weights of 1–1.5 g

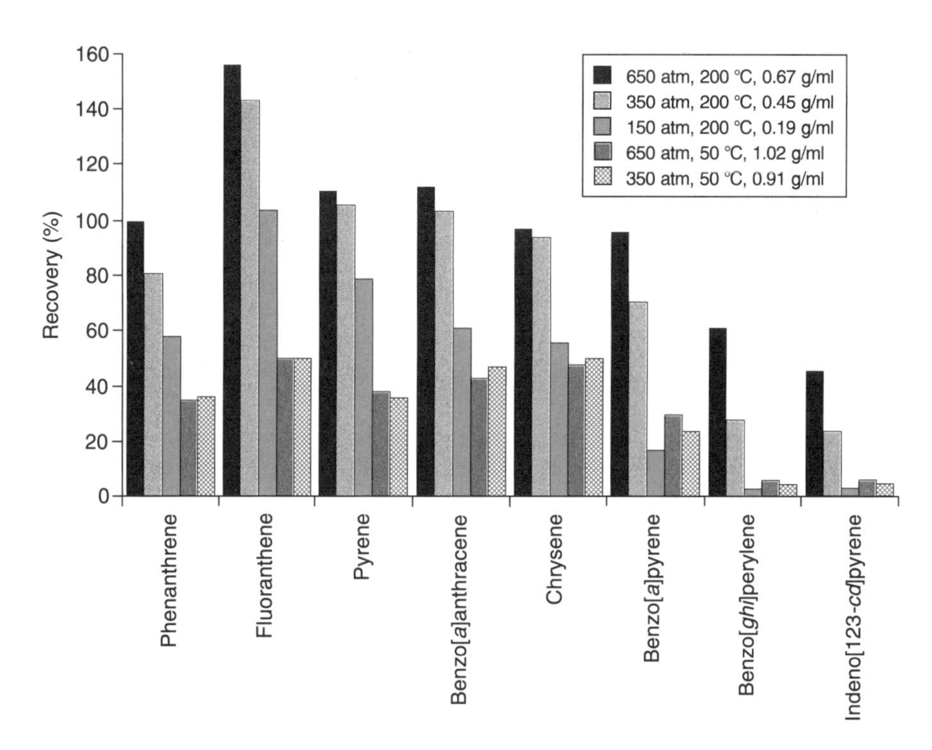

Figure 8.3 SFE of PAHs from urban air particulates (SRM CRM 1649). From Langenfeld *et al.*, *Analytical Chemistry*, **65** (1993) 338

and 0.4 g were used for the soil and diesel soot, respectively. The effect on the recovery of PAHs, as compared to Soxhlet extraction, was dramatic from low temperature (50 °C) to high temperature (200 °C) using a 30 min extraction. The average recovery of 17 individual PAHs from the railroad soil was 56% at 50 °C, 81% at 200 °C and 90% at 350 °C whereas for 13 PAHs from the diesel soot it was 51% at 50 °C, 71% at 200 °C and 118% at 350 °C. The smaller molecular weight PAHs, e.g. naphthalene, were always obtained at higher recoveries than higher molecular weight PAHs, e.g. benzo[*ghi*]perylene. It was generally concluded that higher temperature was favourable for the increased recovery of PAHs from native samples. The use of a temperature of 350 °C was not, however, recommended because it was thought that production of some low molecular weight PAHs may have occurred in the case of the diesel soot. The use of a 30 min extraction, at the increased temperature of 200 °C with carbon dioxide only was considered to produce results which were in agreement with those obtained using a 1 h Soxhlet extraction, thus precluding the addition of an organic modifier.

Subsequent work from the same group[22] investigated the combined effects of temperature and organic modifier on the recovery of PAHs from environmental samples. In this work marine sediment (SRM CRM 1941), diesel soot and air particulate matter (SRM CRM 1649) were extracted with 400 atm of carbon dioxide only or modified carbon dioxide (10% methanol, diethylamine or toluene, added directly to the sample) at temperatures of 80 and 200 °C and a combined extraction time of 30 min (15 min static followed by 15 min dynamic extraction). The results for the two CRMs were compared with certificate values. In the case of CRM 1941 the certificate results were obtained after two sequential 16-hour Soxhlet extractions whereas for CRM 1649 the results were obtained after a 48-hour Soxhlet extraction. The diesel soot sample (0.1 g) was Soxhlet extracted using 150 ml of DCM for 14 h. The results shown in Figures 8.4 and 8.5 for SRM CRM 1941, Figures 8.6 and 8.7 for SRM CRM 1649 and Figures 8.8 and 8.9 for diesel soot indicate that the highest SFE recoveries are obtainable at the higher extraction temperature (200 °C) with diethylamine as the organic modifier. These results were in agreement with those obtained by Soxhlet extraction (14–48 h).

Barnabas *et al.*[23] used a chemometric approach, based on a central composite design, to evaluate the SFE operating parameters of pressure (100–300 kg cm^{-2}), temperature (45–100 °C), extraction time (10–60 min) and percentage methanol modifier (0–20%) on the extraction efficiency of 16 PAHs from native soil. The statistical data concluded that at the 95% confidence level, the operating parameters of importance for the extraction of PAHs from soil were extraction time and percentage methanol modifier. The optimum SFE conditions were concluded to be: pressure, 200 kg cm^{-2}; temperature, 70 °C; extraction time, 60 min; and, a methanol modifier content of 20%. This SFE method was then applied to the contaminated soil in a repeatability study. It was found that the sum of the 16 individual PAHs was 458 mg kg^{-1}. This result compared favourably with that obtained by microwave-assisted extraction (422.9 mg kg^{-1} using acetone as the

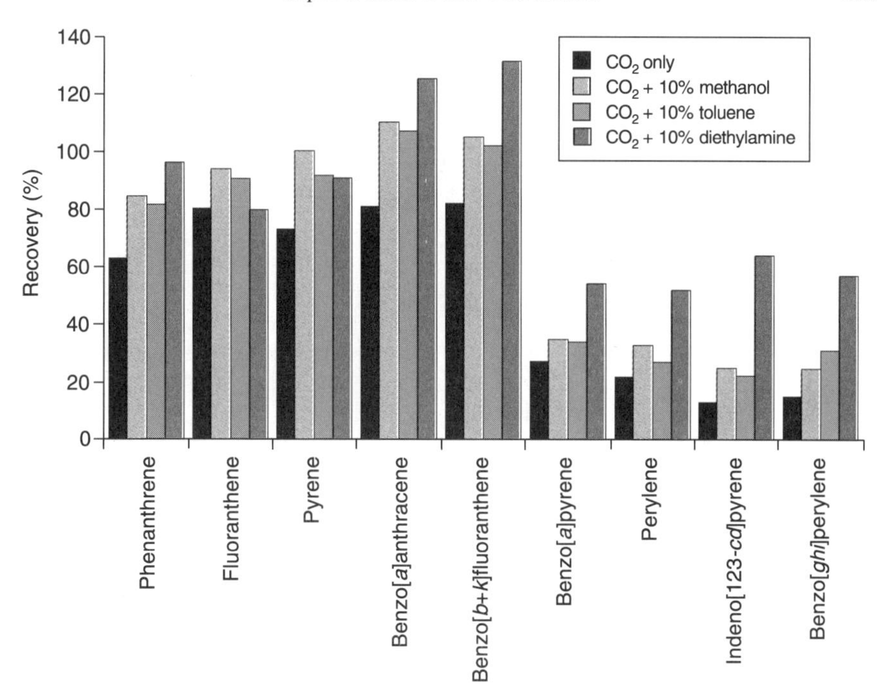

Figure 8.4 SFE of PAHs from marine sediment (SRM CRM 1941) at 80 °C. From Yang *et al.*, *Analytical Chemistry*, **67** (1995) 641

extraction solvent)[24,25] and accelerated solvent extraction (421.9 mg kg^{-1} using acetone and DCM as the extraction solvent)[26] but was superior to that obtained using a 6-hour DCM Soxhlet extraction (297.4 mg kg^{-1}). Results for each individual PAH, by the four extraction techniques, are shown in Figure 8.10. Finally, the method was applied to an interlaboratory soil sample (CONTEST, LGC). In this approach SFE was compared with microwave-assisted extraction and Soxhlet extraction. The SFE results (sum of 16 individual PAHs, 280.9 mg kg^{-1}) compared favourably with the results obtained by MAE (241.7 mg kg^{-1} using acetone and 247.5 mg kg^{-1} using DCM) and Soxhlet extraction (242.7 mg kg^{-1} using acetone). Results for each individual PAH, by the four extraction techniques, are shown in Figure 8.11. It was, however, noted that some blockages did occur during the SFE of the CONTEST sample. Upon investigation it was found that a significant content of elemental sulfur was present in the soil sample. The presence of elemental sulfur caused intermittent blockages for the SFE system. It has been reported previously that the addition of copper to the sample extraction cell (on the exit side) can alleviate this problem.[27] No results were obtained for this CONTEST soil using accelerated solvent extraction as the presence of the sulfur-containing

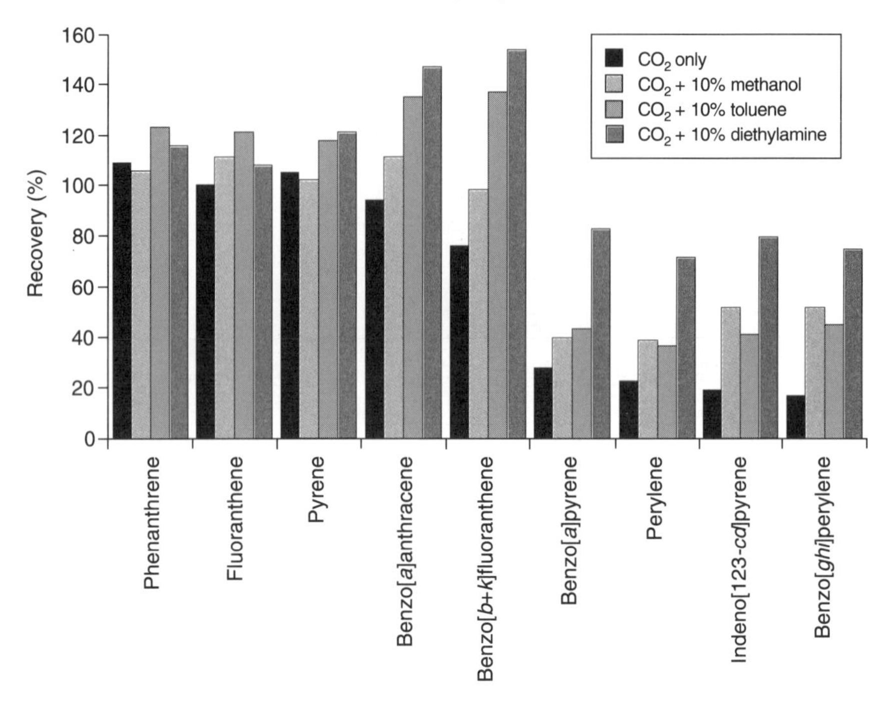

Figure 8.5 SFE of PAHs from marine sediment (SRM CRM 1941) at 200 °C. From Yang *et al.*, *Analytical Chemistry*, **67** (1995) 641

soil caused blockages. It is presumed that the addition of copper to the extraction cell of ASE will also alleviate the problem.

An interesting diversion from the use of supercritical carbon dioxide was that reported by Hawthorne *et al.* who used sub- and supercritical water to extract PAHs from an urban air particulate CRM (SRM 1649).[28] The results (Figure 8.12) indicate that water at 250 °C and 50 bar is an effective media for the extraction of the selected PAHs. These results are compared with the use of supercritical carbon dioxide only (200 °C and 659 bar) and 10% toluene-modified supercritical carbon dioxide at 80 °C and 405 bar. It is again noted that poor recoveries are obtained for the higher molecular weight PAHs.

8.3.2 POLYCHLORINATED BIPHENYLS

Two CRMs (a NIST river sediment, SRM 1939 and a sewage sludge, SRM 392) were analysed for PCBs using a commercial SFE system by David *et al.*[29] The SFE system was optimised for PCBs' extraction using CO_2 only (temperature, 60 °C; density 0.75 g/ml and a 1 ml min^{-1} flow rate of liquid CO_2). The results for the

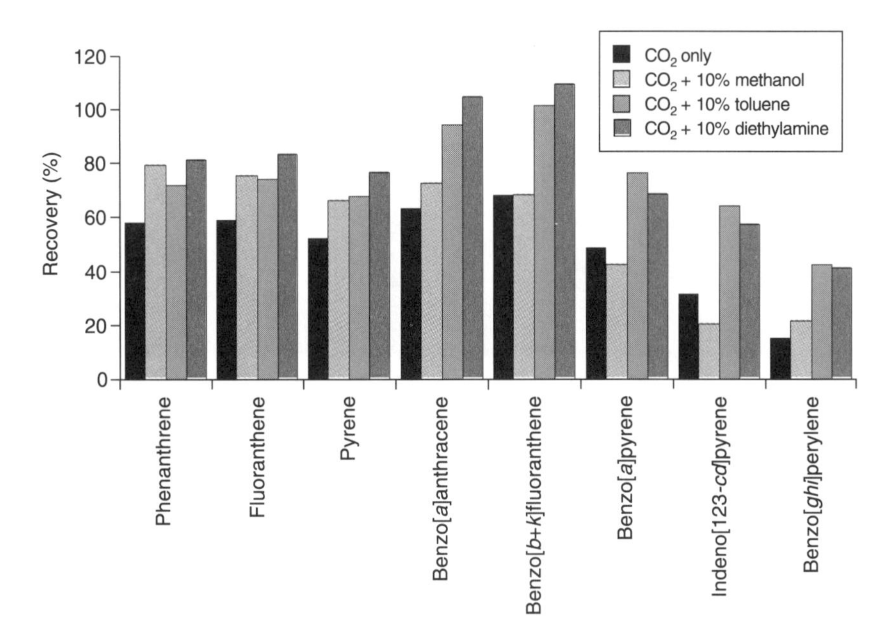

Figure 8.6 SFE of PAHs from air particular matter (SRM CRM 1649) at 80 °C. From Yang *et al.*, *Analytical Chemistry*, **67** (1995) 641

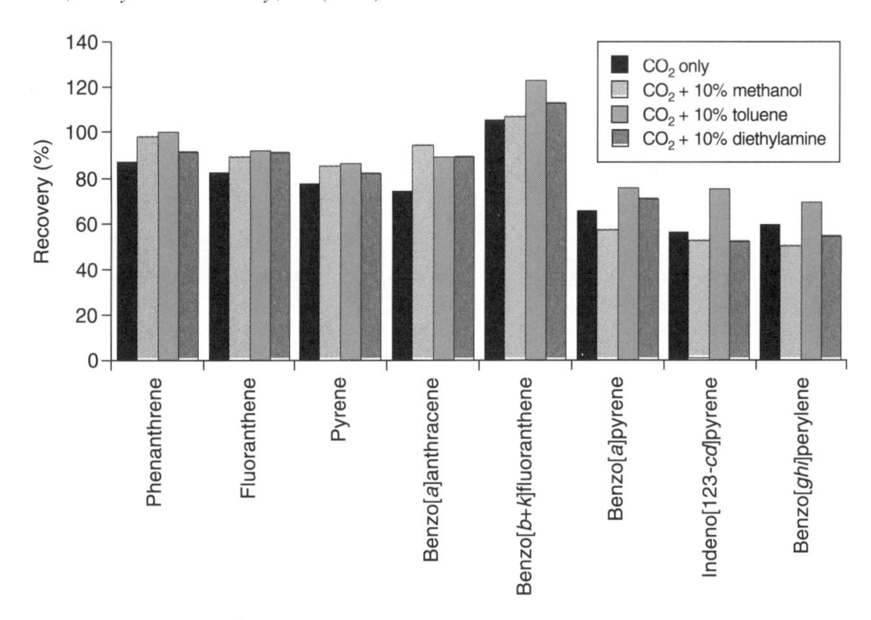

Figure 8.7 SFE of PAHs from air particulate matter (SRM CRM 1649) at 200 °C. From Yang *et al.*, *Analytical Chemistry*, **67** (1995) 641

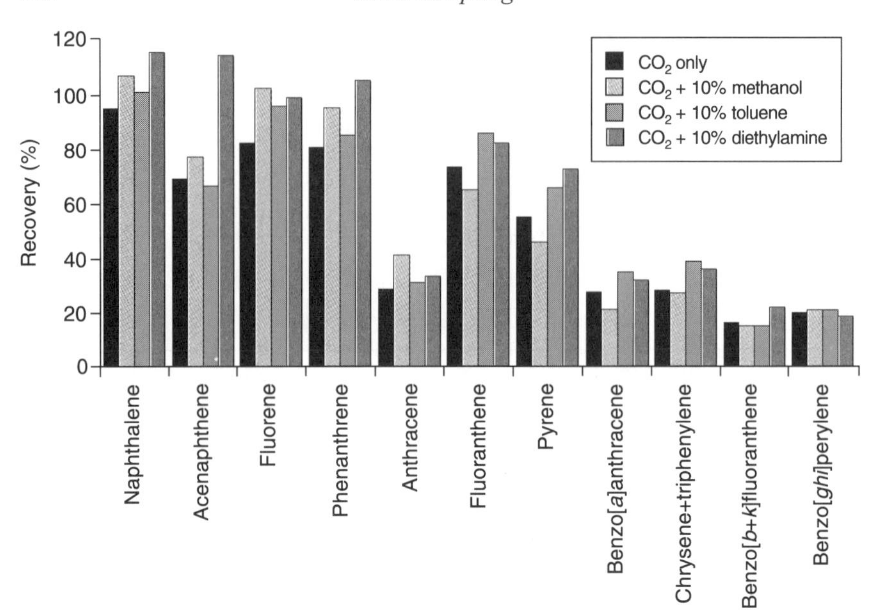

Figure 8.8 SFE of PAHs from diesel soot at 80 °C. From Yang *et al.*, *Analytical Chemistry*, **67** (1995) 641

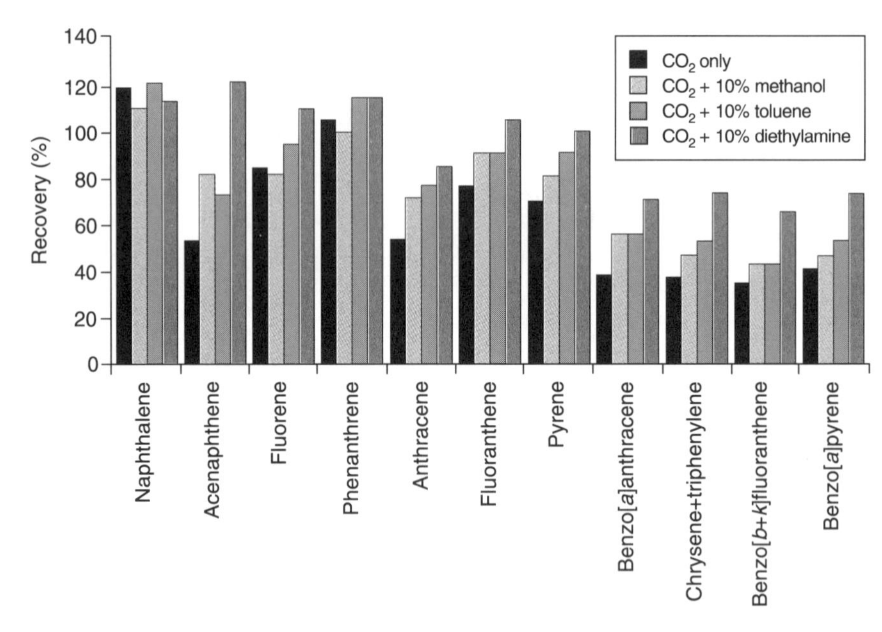

Figure 8.9 SFE of PAHs from diesel soot at 200 °C. From Yang *et al.*, *Analytical Chemistry*, **67** (1995) 641

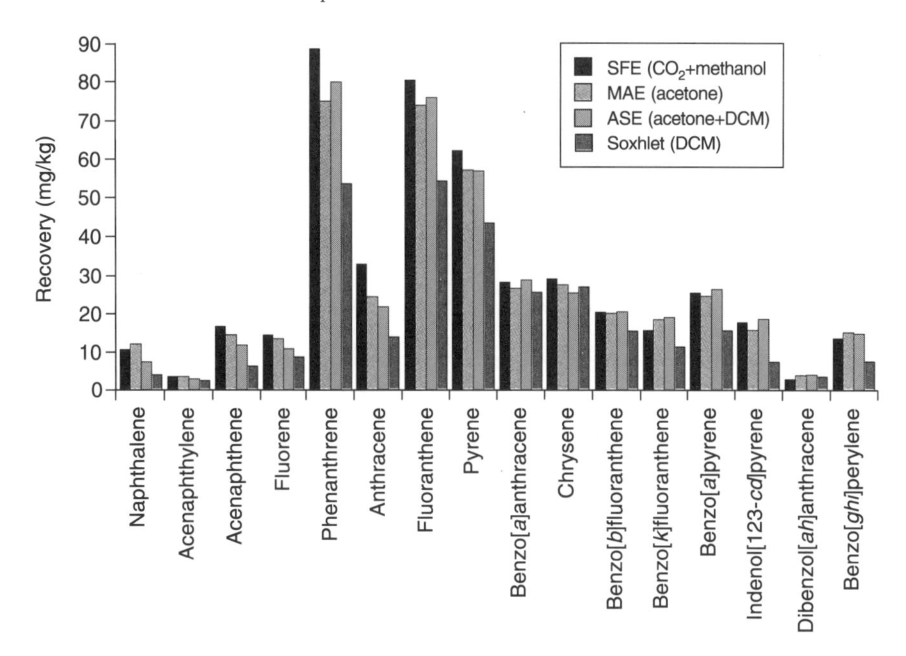

Figure 8.10 Comparison of extraction techniques: extraction of PAHs from contaminated land[23,24,26]

PCB congeners are shown in Figures 8.13 and 8.14 for SRM 1939 and CRM 392, respectively. It is seen that quantitative recovery versus the certificate values are obtained in most cases.

As was mentioned previously[27] the presence of sulfur can cause problems in terms of irreproducibility of SFE and restrictor blockages. One approach to circumvent these problems is to mix copper with the sample prior to addition to the extraction cell. This approach has been utilised by Bowadt and Johansson for the SFE of PCBs from sulfur-containing sediment samples.[30] The SFE conditions used were as follows: 20 min static extraction with pure CO_2 at a density of 0.75 g ml^{-1} (218 atm) at 60 °C followed by a 40 min dynamic extraction under the same conditions. In addition to copper being mixed with the sample (2 g sediment plus 1.5 g copper) 6.5 g of anhydrous sodium sulphate was also added. The addition of anhydrous sodium sulfate ensures a better distribution of the extraction fluid and also enlarges the surface area of the sample. The results from SFE were compared with those obtained by Soxhlet extraction (2 g samples were mixed with 8 g of anhydrous sodium sulfate and extracted with 250 ml of a 2 : 3 mixture of n-hexane and acetone). The results are shown for three sediment types, high- (2.4%) (Figure 8.15), medium- (1.4%) (Figure 8.16) and low-sulfur (1.0%) (Figure 8.17) content obtained as part of a survey of surface sediments from Venice, Italy. The same SFE

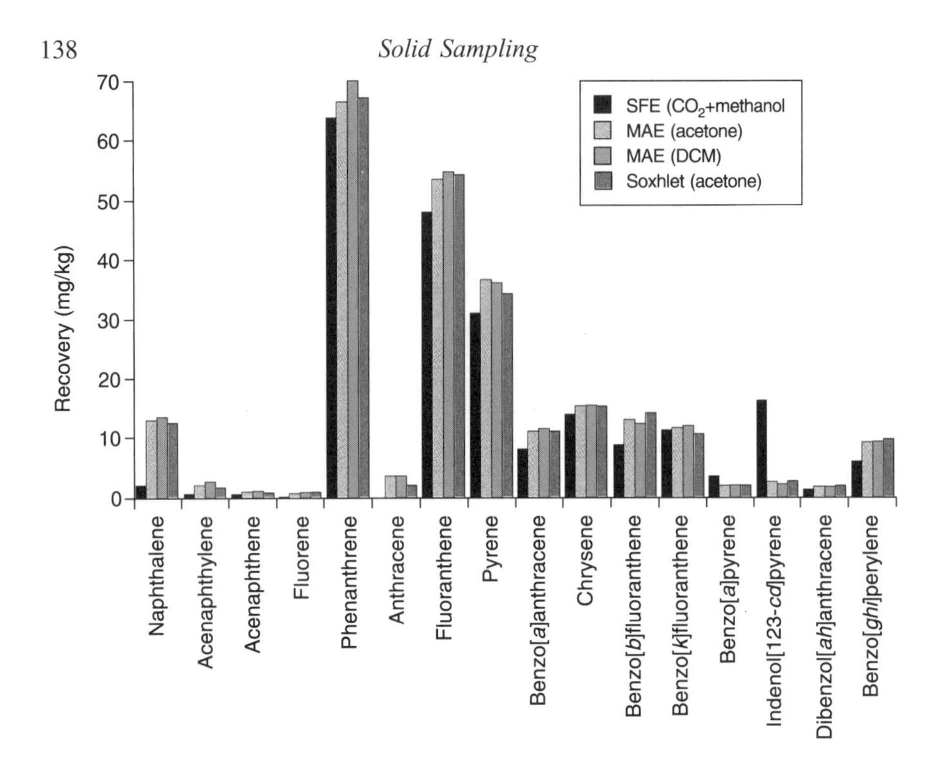

Figure 8.11 Comparison of extraction techniques: extraction of PAHs from an LGC CONTEST soil[23,24]

approach was also applied to the analysis of a CRM (sewage sludge, CRM 392). It is seen that the results obtained by SFE are in agreement with the certificate values (Figure 8.18).

An independent comparison between SFE and Soxhlet extraction was conducted as part of a certification of PCB congeners in industrial soil and reported by Bowadt *et al.*[31] This study was performed in the framework of the PCB group for the Measurements and Testing Programme under the Commission of the European Union. The work involved 21 selected and independent laboratories experienced in congener-specific PCB analysis. All SFE experiments were done on commercial instruments as supplied by a sole manufacturer. Three independent laboratories were involved with SFE. As each laboratory was independent the methods used varied slightly. These different approaches are summarised in Table 8.4.

Figure 8.19 shows the SFE results for the PCB congeners as obtained by the three laboratories using SFE. It is observed that the results obtained are in agreement for each laboratory. A summary of the interlaboratory comparison between SFE and Soxhlet extraction is shown in Figure 8.20. The authors concluded that no further clean-up was required after SFE and prior to analysis by

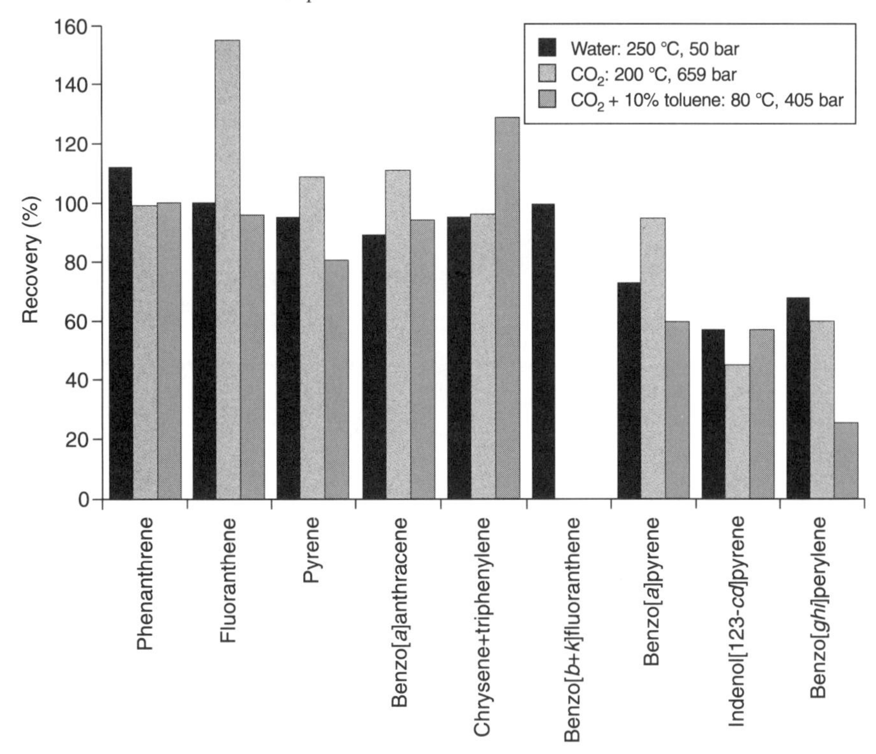

Figure 8.12 SFE of PAHs from urban air particulates (SRM CRM 1649). From Hawthorne *et al.*, *Analytical Chemistry*, **66** (1994) 2912

GC with ECD. Also, that an optimised SFE approach is able to yield extraction with sufficient accuracy and precision to be a significant alternative to conventional methods for the analysis of PCBs in real soil samples, and also for certification purposes.

An alternative to the use of supercritical CO_2 was that reported by Yang *et al.* who used subcritical water to extract PCBs from an industrial soil (CRM 481) and a river sediment (NIST, CRM 1939).[32] The results (Figure 8.21) indicate that water at 250 °C and 50 atm is an effective media for the extraction of the selected PCBs.

8.3.3 PHENOLS

An in situ SFE and derivatisation procedure has been reported for the determination of pentachlorophenol and related compounds from soil samples.[33] In this work phenols are extracted from soil and are acetylated in situ with supercritical CO_2 in the presence of triethylamine and acetic anhydride at a temperature of 80 °C. The

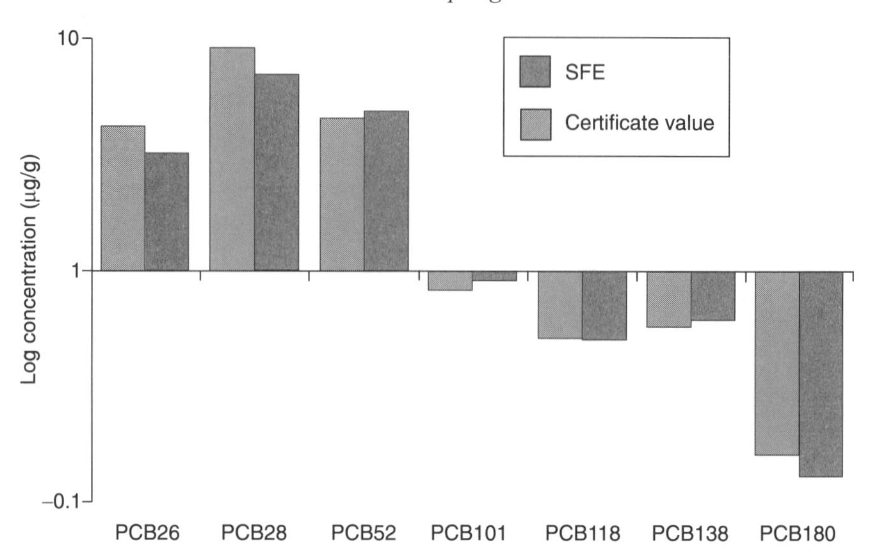

Figure 8.13 Analysis of a river sediment (SRM CRM 1939). From *Fresenius Journal of Analytical Chemistry*, **344** (1992) 479

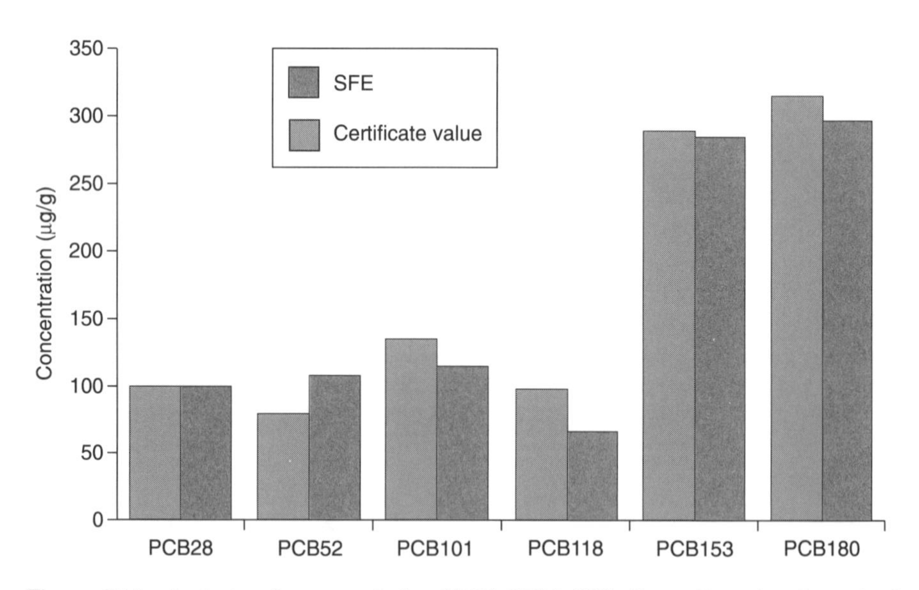

Figure 8.14 Analysis of sewage sludge (SRM CRM 392). From *Fresenius Journal of Analytical Chemistry*, **344** (1992) 479

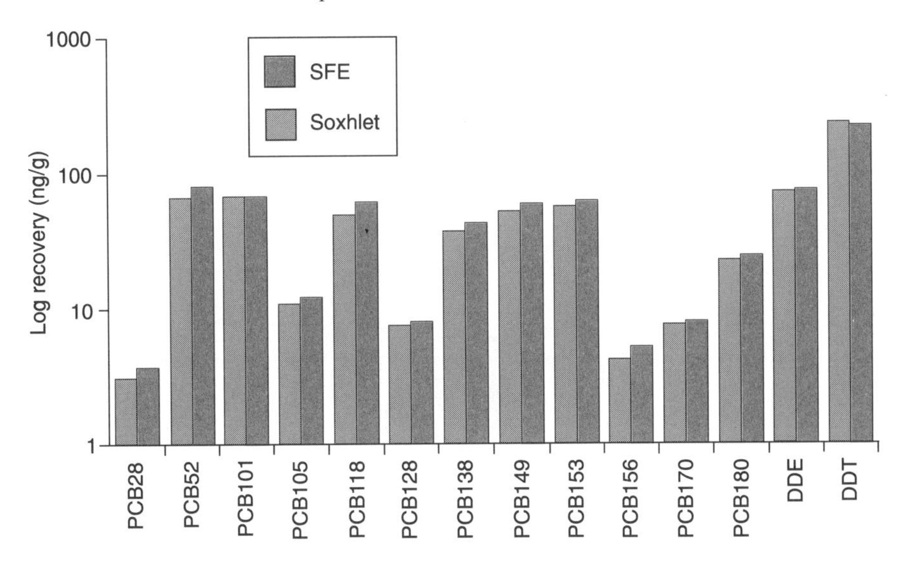

Figure 8.15 Comparison of Soxhlet and SFE: high sulphur content (2.4%). From Bowadt and Johansson, *Analytical Chemistry*, **66** (1994) 667

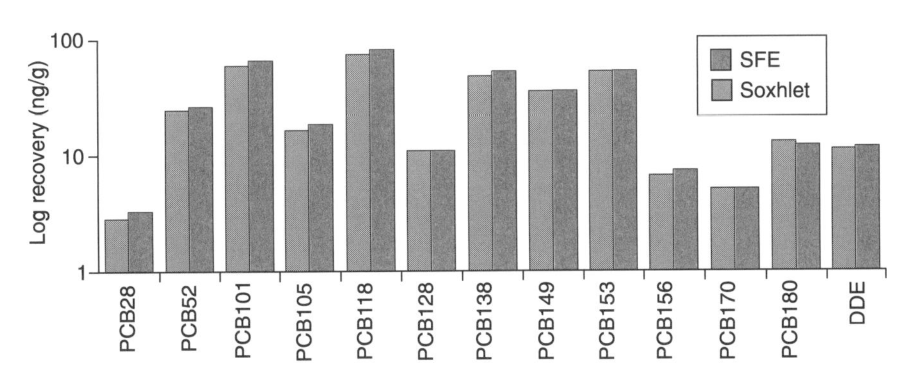

Figure 8.16 Comparison of Soxhlet and SFE: medium sulphur content (1.4%). From Bowadt and Johansson, *Analytical Chemistry*, **66** (1994) 667

method was compared with a steam distillation procedure for the extraction of pentachlorophenol, 2,3,5-trichorophenol, 2,3,5,6-tetrachorophenol, 2,3,4,6-tetrachorophenol and 2,3,4,5-tetrachorophenol from a reference sample (SRS 103-100). The results show similar accuracy to each other (Figure 8.22). The authors conclude that SFE offers a viable alternative to steam distillation.

Hawthorne and Miller have directly compared Soxhlet extraction with low- and high-temperature supercritical CO_2 extraction for the removal of (8) chlorophenols

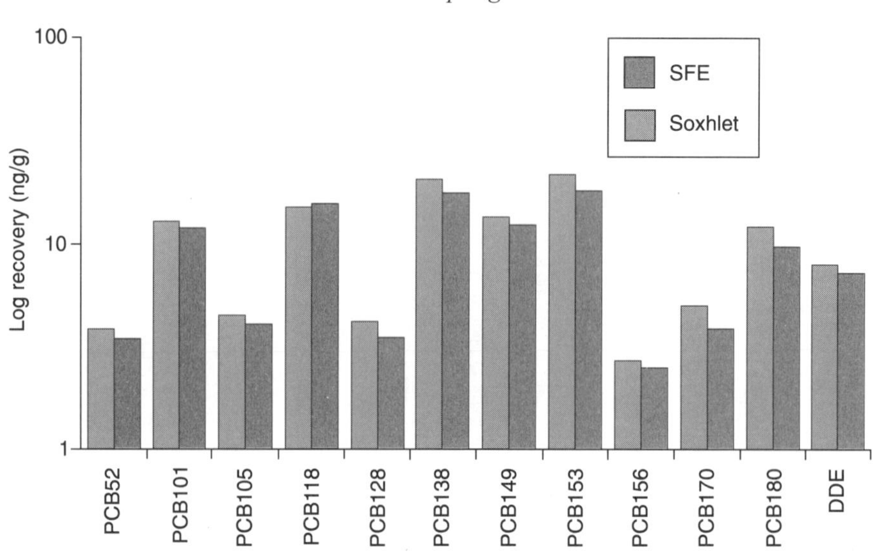

Figure 8.17 Comparison of Soxhlet and SFE: low sulphur content (1.0%). From Bowadt and Johansson, *Analytical Chemistry*, **66** (1994) 667

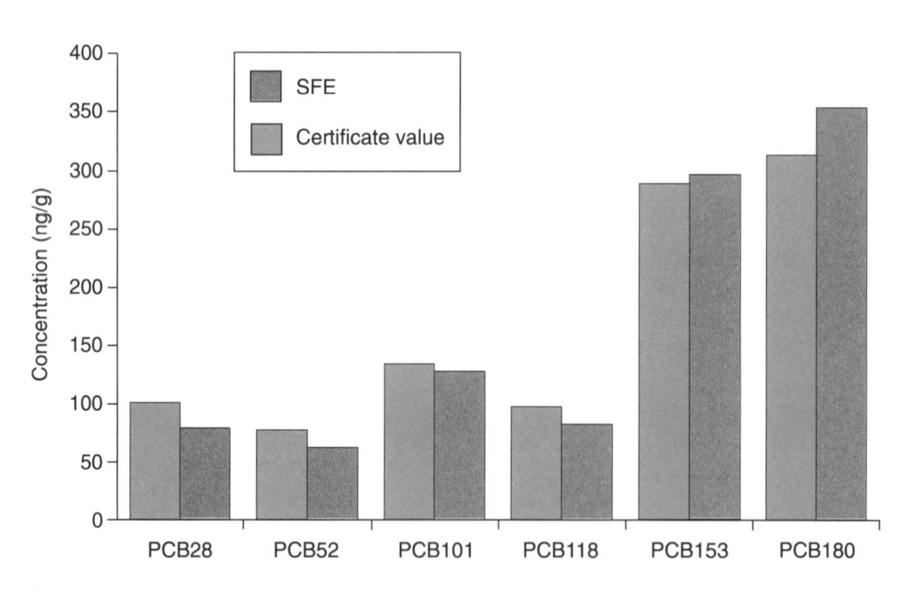

Figure 8.18 Analysis of sewage sludge (SRM CRM 392). From Bowadt and Johansson, *Analytical Chemistry*, **66** (1994) 667

Table 8.4 Procedures for the SFE of PCB congeners from a contaminated soil (CRM 481)[31]

	Laboratory 1	Laboratory 2	Laboratory 3
Sample	0.1 g of sample was mixed with ~10 g of anhydrous sodium sulphate	0.1 g of sample was mixed with ~5 ml of anhydrous sodium sulphate which had previously been mixed with 5.5% v/v of methanol	A cotton-wool plug followed by a layer of ~1 g of anhydrous sodium sulphate, 0.6 g of sample, followed a further layer of ~1 g of anhydrous sodium sulphate and finished by a cotton-wool plug
Supercritical fluid	Pre-mixed cylinder of 2% methanol-modified CO_2	CO_2 only	1.7% methanol-modified CO_2 (methanol added via a second pump)
Temperature (°C)	97	70	70
Pressure (bar)	378	371	371
Density ($g\,ml^{-1}$)	0.75	0.84	0.84
Extraction time	10 min static extraction followed by 40 min dynamic extraction	16 min static extraction followed by 30 min dynamic extraction	20 min static extraction followed by 20 min dynamic extraction
Flow rate (liquid CO_2) ($ml\,min^{-1}$)	1.0	3.0	3.0
Solid phase trapping after depressurisation	~1 ml of Florisil (0.16–0.25 mm particle size)	~1 ml of ODS (Hypersil, 35–45 μm)	~0.3 g of Kieselgel KG60
Elution from solid phase trap	2 × 1.5 ml of n-heptane; 1 × 1.5 ml of DCM; and 2 × 1.5 ml of n-heptane	7 × 1.5 ml of isooctane	8 × 1.0 ml of n-hexane

from an industrial soil.[34] The Soxhlet extraction (18 h) was done using DCM as the extraction solvent. The soil was mixed 1 : 1 with sodium sulphate. For the SFE sample weights of 1–1.5 g were used. The effect on the recovery of the chlorophenols, as compared to Soxhlet extraction, were dramatic from low temperature (50 °C) to high temperature (200 °C) using a 30 min extraction (Figure 8.23). No merit was found in increasing the temperature to 350 °C.

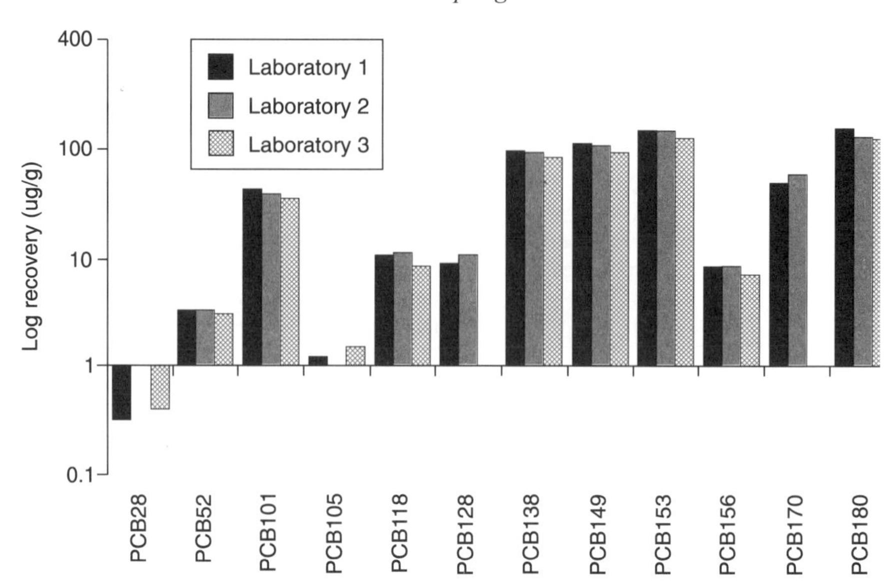

Figure 8.19 Interlaboratory comparison: SFE. From Bowadt *et al.*, *Analytical Chemistry*, **67** (1995) 2424

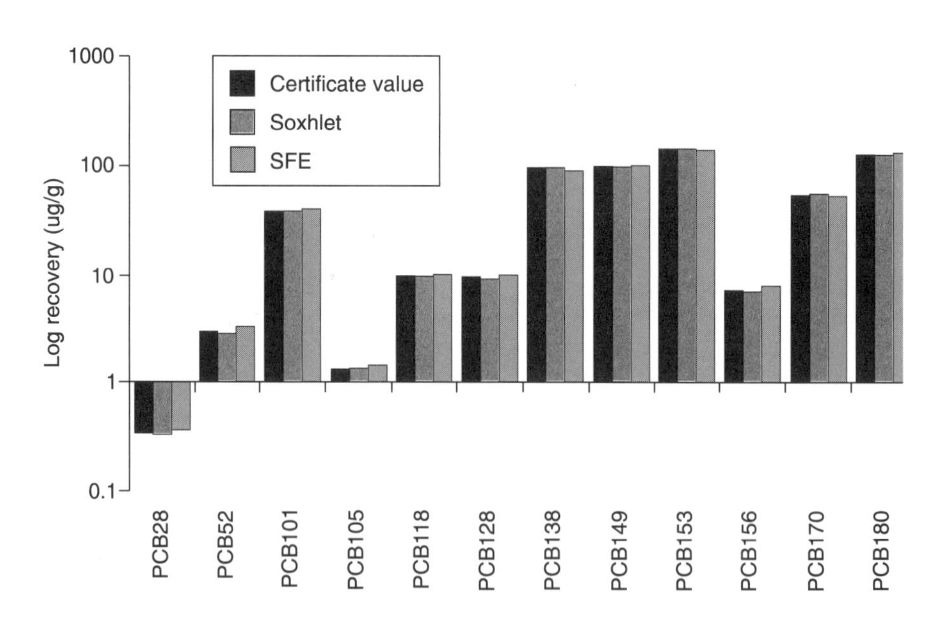

Figure 8.20 Independent laboratory comparison: SFE and Soxhlet. From Bowadt *et al.*, *Analytical Chemistry*, **67** (1995) 2424

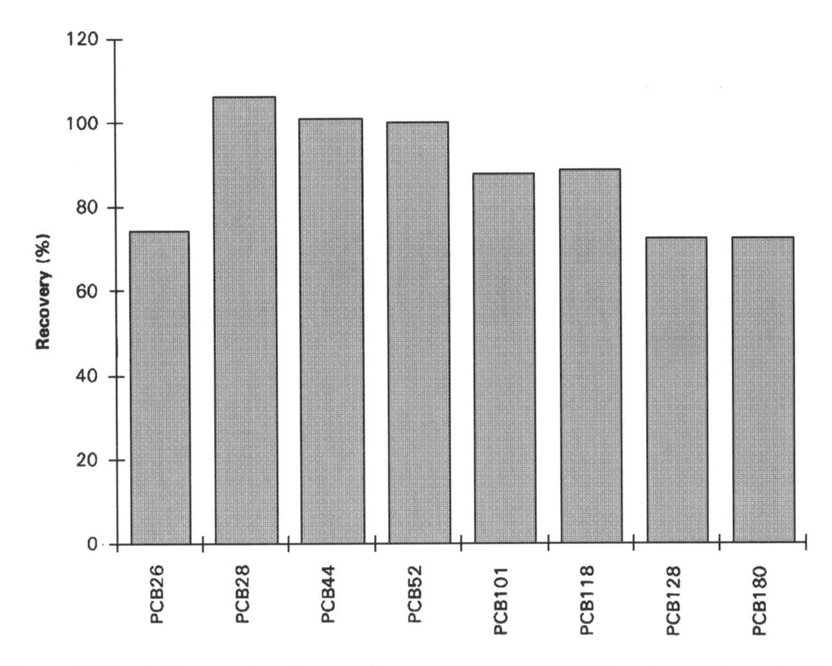

Figure 8.21 PCB extraction from sediment (SRM CRM 1939) using water extraction at 250 °C, 15 atm for 15 min. From Yang *et al.*, *Analytical Chemistry*, **67** (1995) 4571

A multivariate approach to optimisation was applied by Llompart *et al.* for the supercritical derivatisation and extraction of phenol from soil samples.[35] Using a two-level (Packett-Burman) and a three-level (central composite) orthogonal factor design the effects of nine variables were considered: CO_2 flow rate, fluid density, temperature, static extraction time, nozzle and trap temperatures, amount of derivatising agent, pyridine concentration, and time of contact between the derivatising agent and the sample prior to extraction. It was concluded that only extraction cell temperature and the amount of derivatising agent were statistically significant. The optimum conditions for the SFE of phenol were therefore concluded to be: CO_2 flow rate, 1.2 ml min^{-1}; fluid density, 0.4 g ml^{-1}; temperature, 115 °C; static extraction time, 5 min; nozzle and trap temperatures of 45 and 20 °C, respectively; amount of derivatising agent, 70 μl of acetic anhydride; pyridine concentration, 20 μl; and, time of contact between the derivatising agent and the sample prior to extraction, 0 min. Using these conditions a certified reference soil (Environmental Research Associates (ERA) soil, lot no. 329) was extracted and analysed and a recovery of $74.2 \pm 7.9\%$ obtained. The authors suspected that the sample was not homogeneous so a portion of the sample was ground to an approximately 60 μm particle size and reanalysed. The recovery obtained was higher, $81.4 \pm 6.2\%$, which is in good agreement with the expected

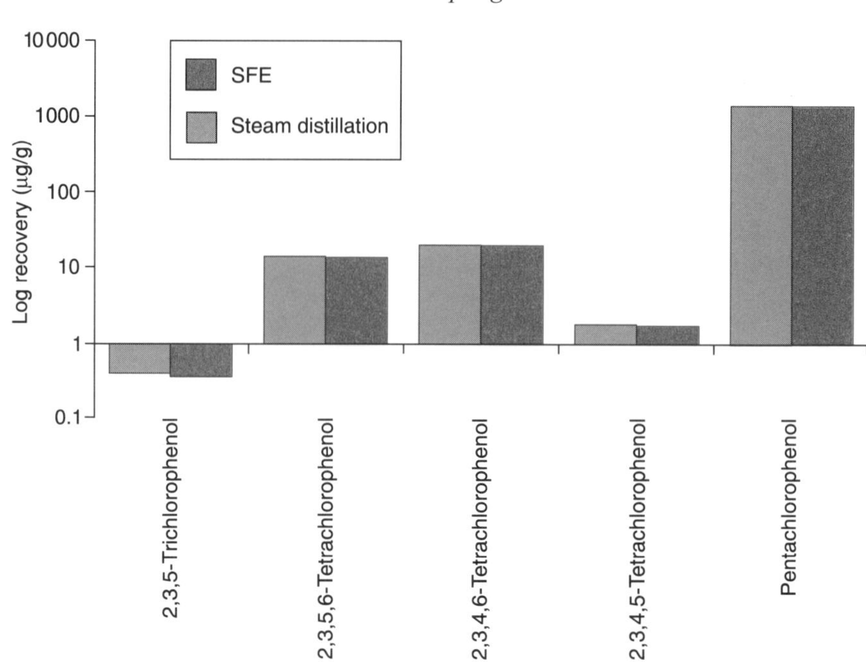

Figure 8.22 Comparison of steam distillation and SFE: reference soil (SRS 103-100). From Lee *et al.*, *Journal of Chromatography*, **605** (1992) 109

value for the extraction procedure. The same authors[36] repeated the work, but instead of derivatising the analytes, they used a methanol-modifier for the SFE of phenol and *o*-, *m*- and *p*-cresols from soil samples. The mean recoveries obtained from the same reference soil (ERA soil, lot no. 329) after grinding to an approximate 60 μm particle size were $80.7 \pm 8.7\%$ for phenol, $17.7 \pm 9.2\%$ for *o*-cresol and 69.2 ± 7.5 for *m*-cresol. No data was presented for *p*-cresol on the reference soil.

8.3.4 PESTICIDES

The literature relating to the SFE of pesticides from environmental matrices is expansive. However, the majority of examples have been done on soils spiked at different concentration levels. While this may provide an introduction to the use of SFE for pesticide analysis it is not appropriate in the development of robust, reliable extraction methods.

A recent review devoted to the effects of soil-pesticide interactions on the efficiency of SFE has been published.[37] In this review two approaches were adopted. The first related to the effect of the soil matrix on pesticide recovery using

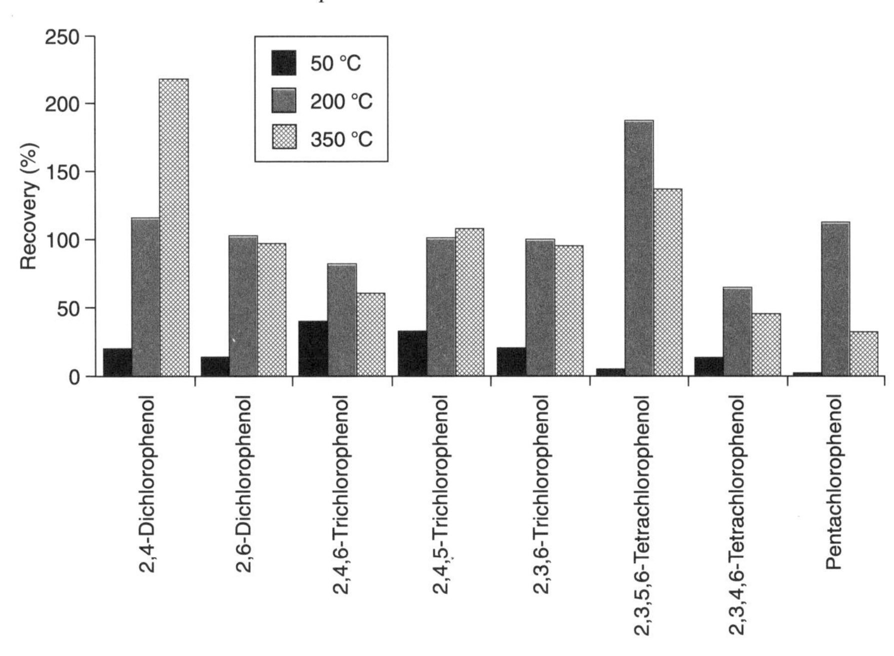

Figure 8.23 Effect of SFE temperature: comparison with an 18 h Soxhlet extraction. From Hawthorne and Miller, *Analytical Chemistry*, **66** (1994) 4005

SFE and secondly on specific examples relating to the SFE of pesticides, herbicides and insecticides. The paper concluded, after suitable review of the literature, by suggesting some important issues to consider for the extraction of pesticides from soils. The main suggestions were:

- Soil characteristics, e.g. organic matter, should be experimentally determined or estimated using appropriate models, and their relationship to pesticide recovery determined. This may allow the development of appropriate extraction models.
- The soil-water partition coefficient (K_d) is indicative of the adsorption of pesticides on soil. Knowledge of K_d may allow appropriate extraction solvents to be selected.
- The solubility of the pesticide in supercritical CO_2 is important.
- Solubility in supercritical CO_2 does not guarantee quantitative extraction.
- Modified supercritical fluids are more likely to overcome matrix effects. The type of modifier to be used should be evaluated; mixed modifiers may be needed.
- Problems may be encountered due to the extraction of co-extractives when using modified supercritical fluids. The use of alternative chromatographic detection systems may be required.
- High pressure leads to high recovery.

- Soils may be washed with a non-polar wash to remove humic substances prior to SFE (care needs to be taken that analytes of interest are not lost).
- Soil organic matter (SOM) appears to be an important parameter affecting analyte recovery. High SOM often leading to poor recovery of analytes.
- In situ derivatisation may lead to improved recovery.

To emphasise the importance of SFE on the recovery of pesticides from native environmental matrices selected examples will now be used. Snyder *et al.* compared sonication with SFE for the extraction of pesticides from native samples.[38] The SFE was operated as follows: pressure, 350 atm; temperature, 50 °C; 3% methanol-modified CO_2 (using a premixed cylinder); and, a static extraction time of 10 min followed by 20 min dynamic extraction using the 2 ml extraction cell. Sonication was done according to the EPA sonication method and briefly comprises extracting 10 g samples for 3 min with 3 × 40 ml portions of 1 : 1 DCM : acetone. The extracts were combined and reduced in volume to 1–2 ml prior to dilution to 5 ml. The results for the three natively contaminated soils are shown in Figures 8.24–8.26. It can be seen that the results obtained by SFE are in agreement with those obtained by sonication.

In an attempt to establish the influence of soil characteristics on the recovery of herbicides Steinheimer *et al.* fortified western cornbelt soils prior to SFE.[39] Four different soil matrices were obtained and their properties characterised using standard procedures (Table 8.5). The soils were spiked with the herbicides (atrazine, desethylatrazine, desisopropylatrazine, cyanazine and metolachlor) in the

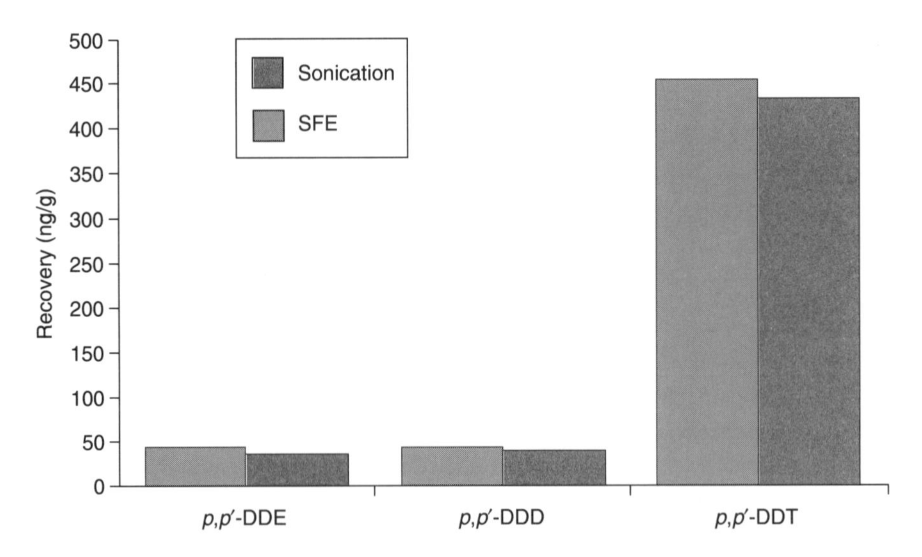

Figure 8.24 Comparison of SFE and sonication: native soil 1. From Snyder *et al.*, *Analytical Chemistry*, **64** (1992) 1940

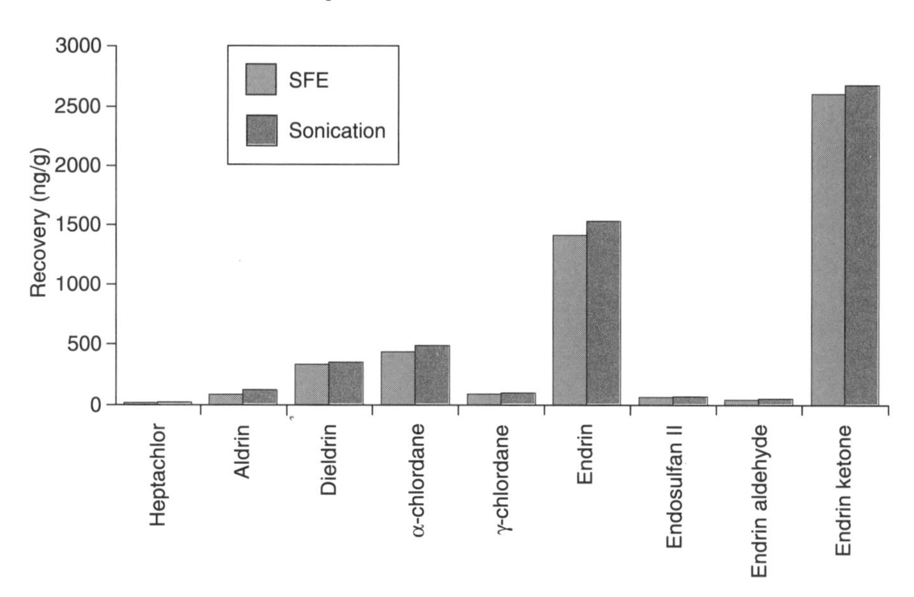

Figure 8.25 Comparison of SFE and sonication: native soil 2. From Snyder *et al.*, *Analytical Chemistry*, **64** (1992) 1940

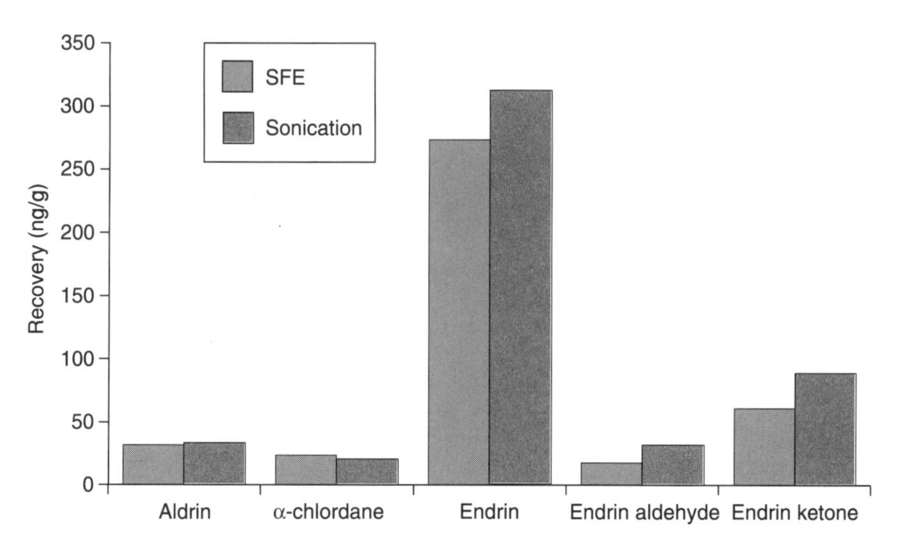

Figure 8.26 Comparison of SFE and sonication: native soil 3. From Snyder *et al.*, *Analytical Chemistry*, **64** (1992) 1940

Table 8.5 Properties of soil samples used for SFE of herbicides[39]

	Kenyon loam	Ida/Monoma silt loam	Synthetic soil matrix	Sand
Organic carbon (%)	4.48	1.81	0.22	< 0.05
Cation exchange	36.10	32.30	10.70	not determined
pH	6.70	6.10	8.30	10.70
Sand (%)	42.50	10.00	54.80	100.00
Silt (%)	47.50	77.50	29.90	0.00
Clay (%)	10.00	12.50	13.80	0.00

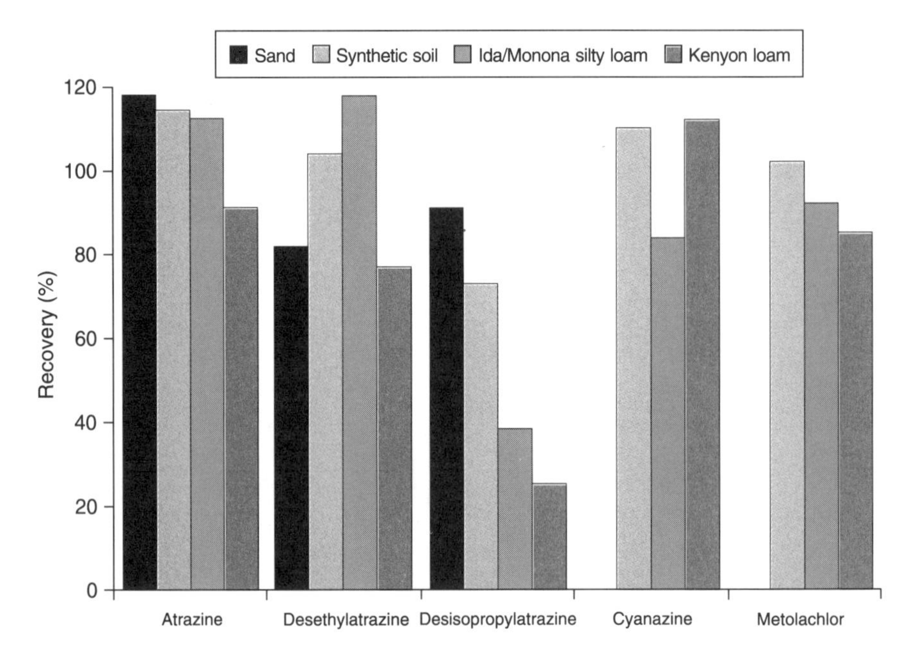

Figure 8.27 SFE of herbicides from fortified western cornbelt soils. From Steinheimer *et al.*, *Analytical Chemistry*, **66** (1994) 645

range 0.1–2.0 mg kg^{-1}. In order to optimise the SFE operating conditions a principal components approach was used. The results concluded that for the SFE of atrazine and its metabolites (desethylatrazine and desisopropylatrazine) high extraction pressure and low temperature, using a maximum of 4–5% water as modifier (cosolvent) was required. For cyanazine and metolachlor, lower extraction pressure is recommended. The results (Figure 8.27) indicate the dependence of the

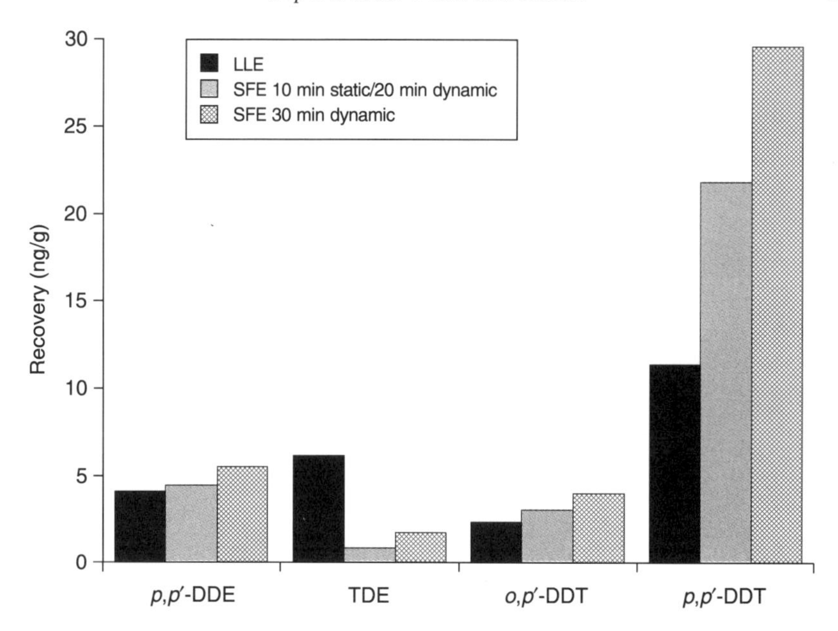

Figure 8.28 Comparison of SFE and LLE: grass land. From van der Velde *et al.*, *Journal of Chromatography*, **683** (1994) 167

atrazine metabolites on soil type. The average results for each herbicide for each soil type are as follows: Kenyon, 72 ± 15%; Ida/Momona, 89 ± 14%; synthetic soil, 96 ± 11%; and, sand, 98 ± 8%.

Similar work has been reported by van der Velde *et al.* for the SFE of organochlorine pesticides (OCPs) spiked in the range 1–10 ng g^{-1} on real soil samples and glass beads.[40] Using SFE conditions of pressure, 20 MPa and a temperature of 50 °C the effects of extraction time were evaluated for three soil types (grassland, agricultural land and orchard soil in Figures 8.28, 8.29 and 8.30, respectively). The results were compared with liquid–solid extraction (shake-flask extraction) and found to be in agreement. The effects of modifier addition (toluene, acetonitrile and methanol) and the influence of the volume of the methanol modifier addition directly to the sample were also investigated. The authors concluded that the development of SFE methods on spiked glass beads offers no relevant information to the parameters required for the extraction of OCPs from spiked soils.

An investigation into the recovery of chlorsulfuron and metsulfuron methyl from four different soils has been investigated.[41] The soils (Gordola, Avully, St Cierges and Courtelle) were collected from sites in Switzerland, sieved to 2 mm, dried and sterilised by heating to 120 °C. The soils were characterised with respect to pH, organic carbon, exchange capacity and clay content. After investigation it was concluded that the best operating conditions for SFE were as follows: temperature,

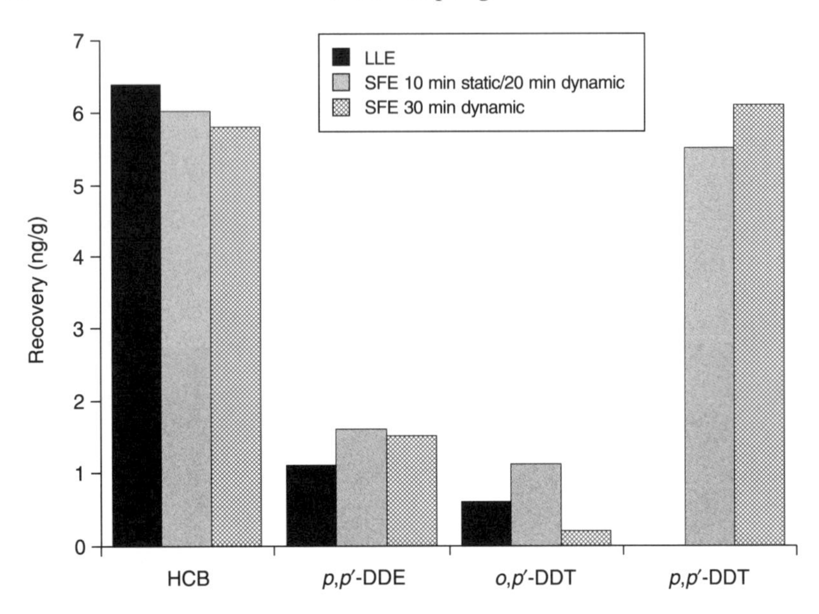

Figure 8.29 Comparison of SFE and LLE: agricultural land. From van der Velde *et al.*, *Journal of Chromatography*, **683** (1994) 167

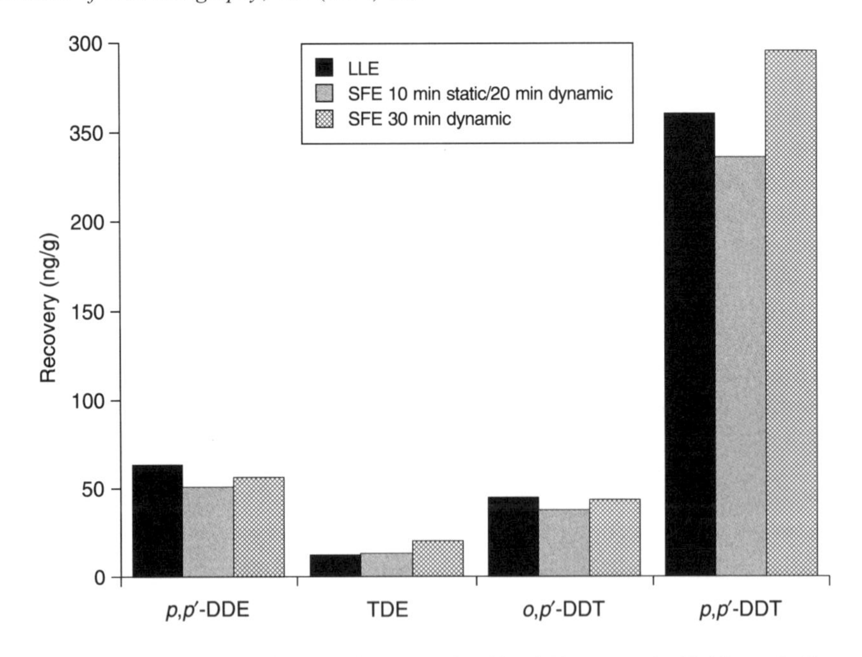

Figure 8.30 Comparison of SFE and LLE: orchard land. From van der Velde *et al.*, *Journal of Chromatography*, **683** (1994) 167

50 °C; pressure, 370 bar; an aliquot (80 μl) of methanol per 4 g soil; and 10 min static extraction followed by 8 min dynamic extraction. Acceptable recoveries (87–90% for chlorsulfuron and 87–91% for metsulfuron methyl at the 5 $\mu g\, g^{1}$ spike level and 82–84% for chlorsulfuron and 75–82% for metsulfuron methyl at the 1 $\mu g\, g^{-1}$ spike level) were obtained for both herbicides from three of the soils, the exception being the soil with the highest organic carbon content (4.2%) and clay content (56%). In this situation only 47–53% of the spiked herbicides could be recovered, irrespective of the spike level.

The effects of the soil matrix on the recovery of organochlorine and organophosphorus pesticides was reported by Snyder *et al.*[42] Four soils (sand, river sediment, clay and top soil) were evaluated plus a furnace-treated (400 °C) soil in which as much organic matter as possible is removed. Spiking involved the addition of a 20 μl spike to the soil (2 g) directly in the extraction cell followed by a time period to allow the solvent (acetone) to evaporate. The SFE was then done using supercritical CO_2 only and with 3% methanol-modified supercritical CO_2. The other SFE conditions were kept constant at 350 atm, 50 °C and an extraction time of 10 min static and 10 min dynamic. Using the 3% methanol-modified CO_2 excellent recoveries were achievable in all cases. However, with CO_2 only some analyte dependence was noted. Perhaps most dramatic was diazinon which was recovered from sand with 81% efficiency but with the other sample matrices was extremely poorly recovered (no recovery was noted at all from the clay sample). Based on this work the authors postulated on the role of modifiers in SFE. Three possibilities were suggested. First, that the addition of modifier allows a more polar solvent system to be formed. Secondly, that the modifier may displace polar analytes from the adsorption sites on the soil, and finally, that the modifier can swell the soil matrix thereby exposing the internal structure of the matrix to the supercritical fluid. It was suggested that the extraction mechanism is a combination of all three.

8.4 METHODS OF ANALYSIS: EXTRACTION FROM AQUEOUS SAMPLES

An interesting application of SFE has been its use for the extraction of analytes from water-based matrices. This has been suggested and applied in two distinct ways. The first involves the direct use of CO_2 to remove analytes from water samples directly while the second involves a combined solid phase extraction-SFE approach. In the latter case, the aqueous sample containing the analytes is first passed through a preconditioned solid phase extraction (SPE) media. Then the SPE media is inserted into the SFE extraction cell and eluted with a suitable supercritical fluid.

These approaches have been applied to a range of sample types of environmental importance and these will be briefly discussed.

8.4.1 DIRECT EXTRACTION OF ANALYTES FROM AQUEOUS SAMPLES

The general principle of this approach is that the aqueous sample is introduced into the extraction cell and then the CO_2 is used to 'purge' the analytes from the cell. Two cell designs have been applied to this approach (Figures 8.31 and 8.32). The first cell type (Figure 8.31) has been applied to the extraction of phosphonate,[43] phenol[44] and a range of organic bases, e.g. triprolidine, sulfamethazine and

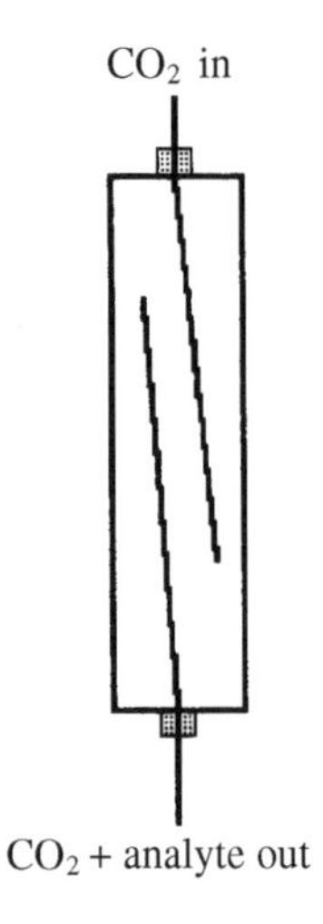

Figure 8.31 Aqueous extraction using a modified extraction cell

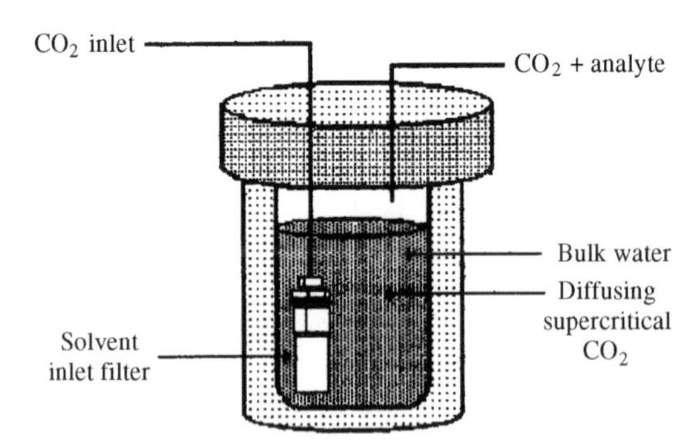

Figure 8.32 Aqueous extraction using a commercially available cell

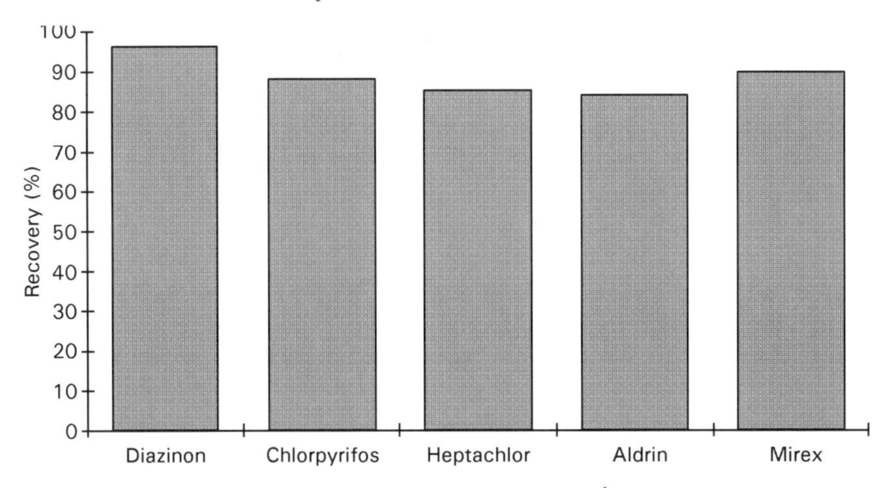

Figure 8.33 SPE-SFE: CO_2 only (spike level 100 ng ml^{-1}). From Ezzell and Richter, *Journal of Microcolumn Separation*, **4** (1992) 319

caffeine,[45] from aqueous samples. In contrast the second cell type (Figure 8.32) has been applied to the extraction of organochlorine pesticides[46] and an alcohol phenyl ethoxylate surfactant[47] from aqueous samples.

8.4.2 USE OF A COMBINED SPE-SFE APPROACH

The basis of this approach is that an aqueous sample is passed through a preconditioned solid phase media. The solid phase media is then inserted into an

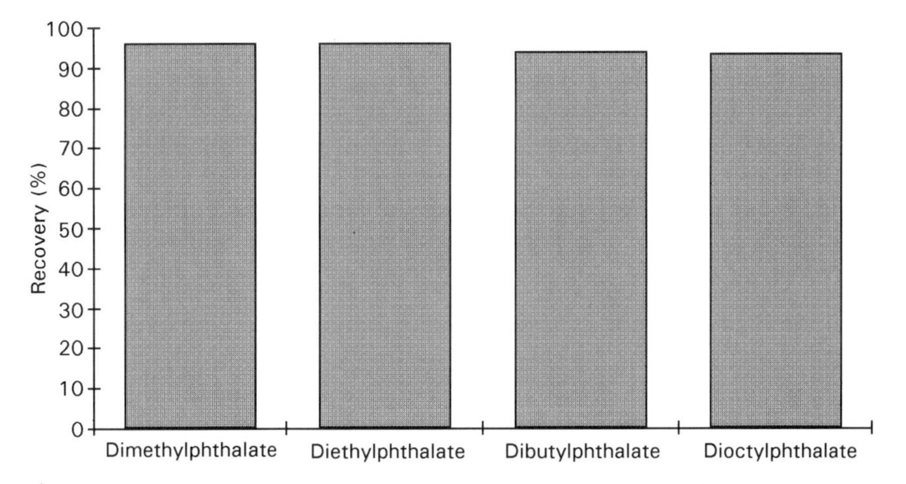

Figure 8.34 SPE-SFE: $CO_2 + 10\%$ methanol (spike level 100 ng ml^{-1}). From Ezzell and Richter, *Journal of Microcolumn Separation*, **4** (1992) 319

extraction cell and eluted with the supercritical elution solvent. This approach has been applied to a variety of situations including: five pesticides (Figure 8.33) and four phthalate esters (Figure 8.34)[48] using a range of SPE media, however, C18 was found to be the best; a selected group of PAHs, PCBs, OCPs and phthalate esters[49] using both C18 SPE cartridges and disks (Figure 8.35); phenols[50] using styrene-divinylbenzene SPE media (Figure 8.36); an alcohol phenyl ethoxylate surfactant using a C18 SPE disk[47] in which the analyte was recovered with an approximate 90% efficiency; 43 semivolatile organic analytes[51] using C18 SPE disk media (Figure 8.37); and 16 PAHs with an average recovery of 99% (RSD 4.6%) using C18 SPE disk media.[52]

A novel approach to this work has been its application towards an investigation of the selectivity of extraction. The aqueous samples are loaded onto SPE media as before and then transferred into the extraction cell of the SFE system. However, the purpose this time is to obtain class selective elution by either a change in supercritical density or by the addition of an organic modifier to the supercritical CO_2. Kane *et al.* were able to demonstrate the effectiveness of this approach by (almost) selectively separating an alcohol ethoxylate from an alcohol phenyl ethoxylate by changing the supercritical fluid density from $0.75\,g\,ml^{-1}$ to $0.85\,g\,ml^{-1}$.[53] Subsequent work from the same group[54] demonstrated the (almost) selective extraction of three organochlorine pesticides from three organophosphorus pesticides. The OCPs could be effectively separated from the C18 SPE disk media using supercritical CO_2 only (13.5 MPa and 50 °C) whereas the OPPs were retained. Addition of methanol-modifier (400 μl) and high pressure (35 MPa) allowed the removal of OPPs. This was followed by the application of selective SPE-SFE to the extraction of OCPs and two classes of herbicides from aqueous samples.[55] In this situation three OCPs could be (almost) selectively eluted from the C18 SPE disk media in preference to the herbicides (triazines and urea herbicides) using the following conditions: pressure, $250\,kg\,cm^{-2}$; temperature, 50 °C; and, supercritical CO_2 only. The OCPs were then analysed using GC-MSD. Addition of 10% methanol-modified supercritical CO_2 allowed the removal of the two classes of herbicides (Figure 8.38). The triazines and urea herbicides were then analysed using HPLC with UV detection. (Note: It should be stressed that the authors did not claim 100% selectivity in any of the examples. However, it seems likely that approximately 90% selectivity is achievable by modification of the supercritical fluid conditions.)

8.5 RECOMMENDATIONS FOR SFE

A procedure for developing a quantitative method for SFE of environmental samples has been proposed by Hawthorne *et al.*[56] and is shown schematically in

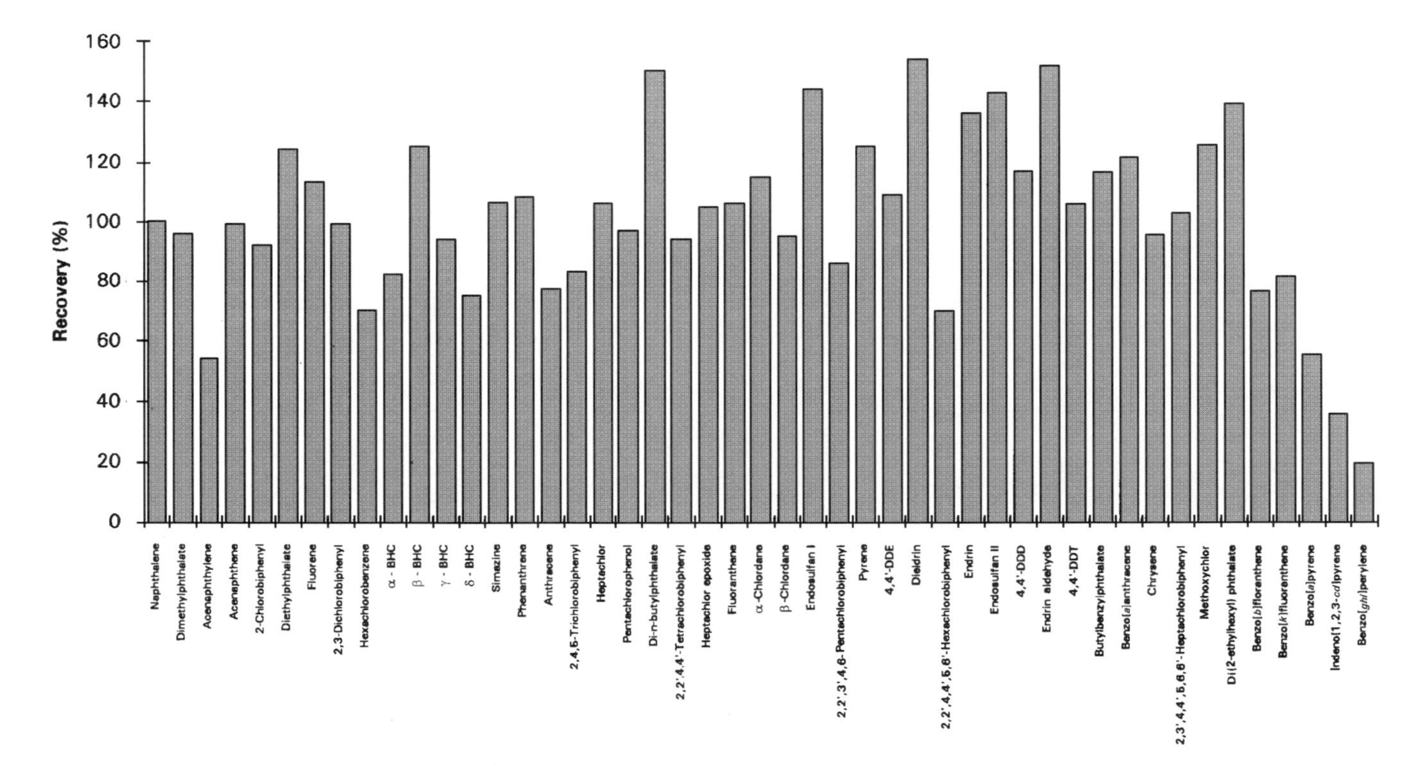

Figure 8.35 SPE-SFE using C18 disk. From Tang *et al.*, *Journal of the Association of Official Agricultural Chemists*, **76** (1993) 72

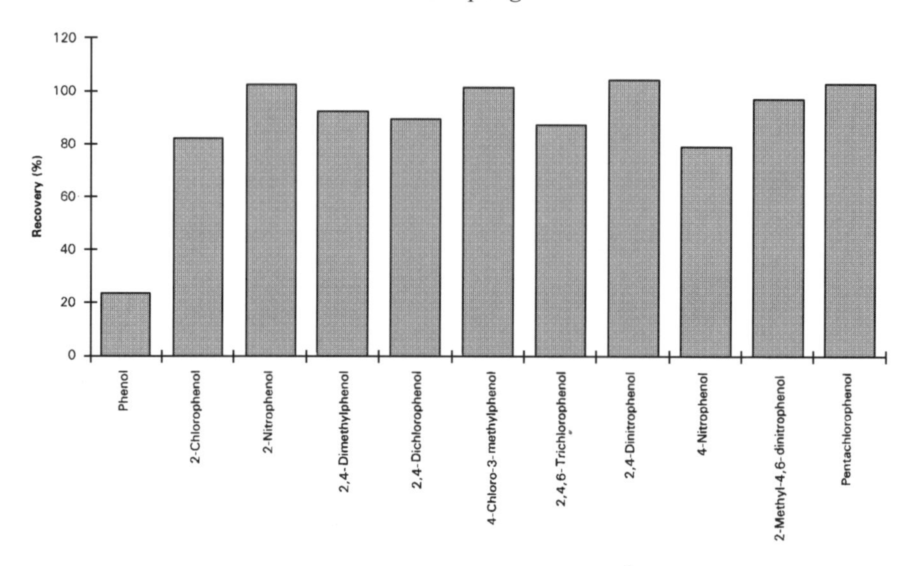

Figure 8.36 SPE-SFE using SDB disks (spike level 1 ng ml^{-1} in 1 litre of reagent water, $n = 7$). From Tang and Ho, *Journal of High Resolution Chromatography*, **17** (1994) 509

Figure 8.39. The procedure is subdivided into various sections and these will now be summarised.

8.5.1 SELECTION OF INITIAL EXTRACTION CONDITIONS

Develop initial extraction conditions based on (a) polarity of target analytes (and solubility in supercritical CO_2, if known), (b) matrix composition (e.g. organic carbon content, water content, particle size, mineral composition) and (c) previously reported successful literature SFE methods. If limited data is available on the solubility or extraction results for the target analyte, initial conditions may be chosen based on the polarity of the analyte, i.e. using the general rule that supercritical CO_2 only will generally solvate GC-able analytes at normal extraction conditions, such as 400 atm and 50 °C. If the analytes are fairly polar or have high molecular masses, then the addition of an organic modifier (10% v/v) may be necessary with an increase in temperature (70 °C). In the case of ionic analytes the addition of an ion-pairing reagent may be beneficial.

8.5.2 PRELIMINARY EXTRACTIONS OF REPRESENTATIVE SAMPLES

In this situation it is important to extract some samples that reflect the range of analyte, water and co-extractable matrix concentrations that may be expected in

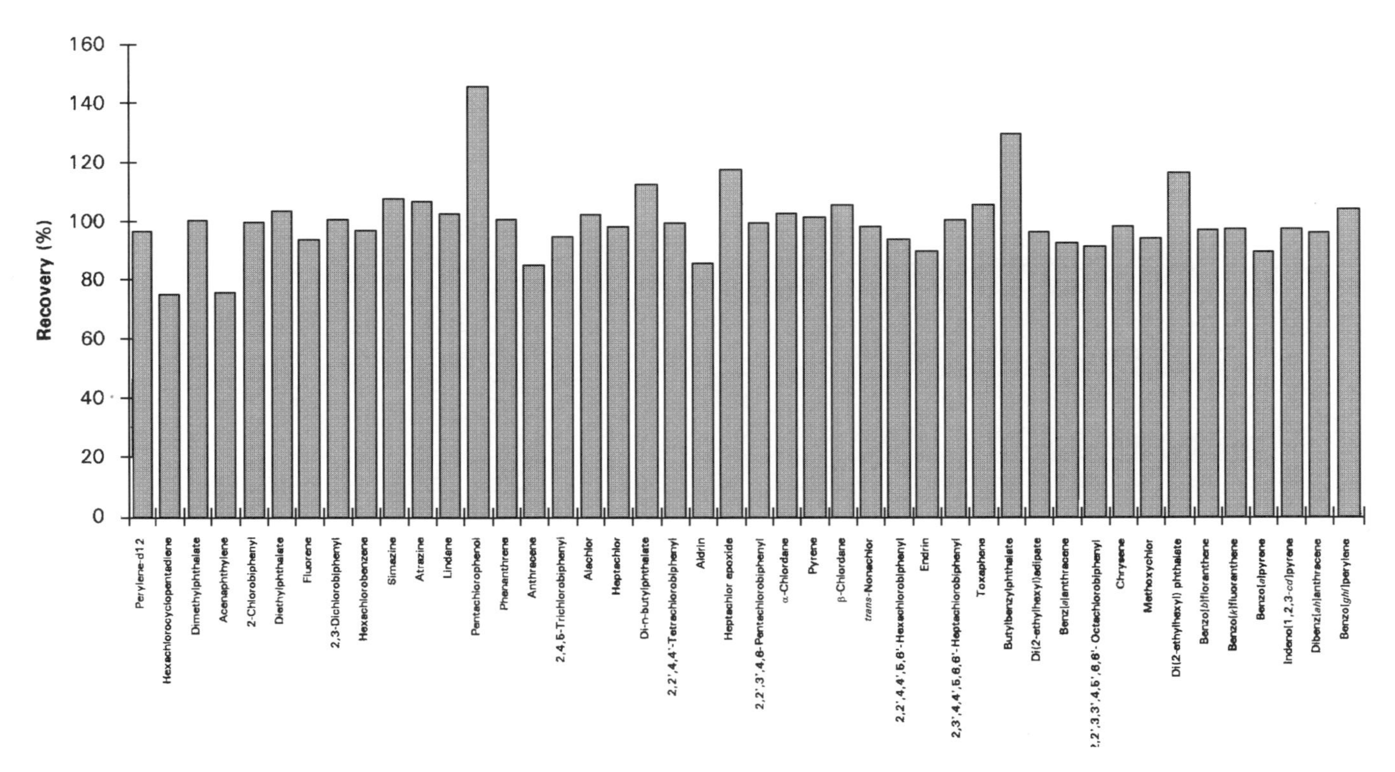

Figure 8.37 SPE-SFE using SDB disks (spike level 2 ng ml^{-1}, $n = 7$). From Tang *et al.*, *Journal of Chromatographic Science*, **33** (1995) 1

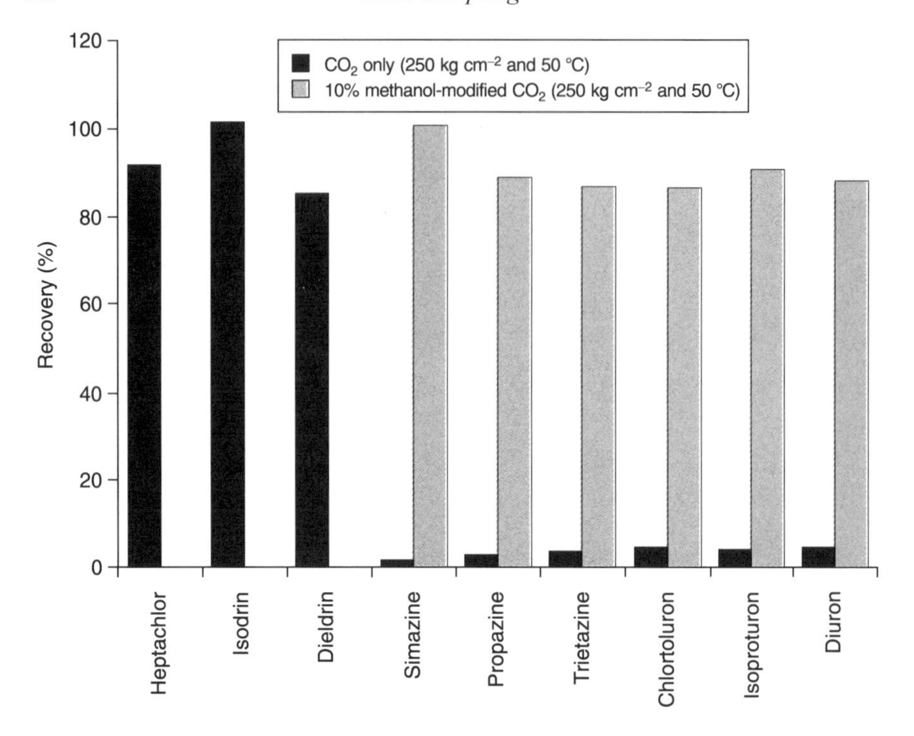

Figure 8.38 Selective SPE-SFE of OCPs and herbicides. From Barnabas *et al.*, *Journal of Chromatographic Science*, **32** (1994) 547

future samples. This will allow some development of a suitable extraction strategy. This may take the form of the mixing of copper with the sample to eliminate potential sulfur interferences and/or the addition of diatomaceous earth or anhydrous sodium sulphate to eliminate moisture effects and provide a large surface area for analyte-matrix interactions.

8.5.3 DETERMINATION OF COLLECTION EFFICIENCIES

Once initial extraction conditions have been evaluated it is important to determine that the target analytes will not be lost after SFE depressurisation. This can be achieved using spiked samples on an inert matrix, e.g. Celite. This will enable an assessment of the collection system. In our laboratory[23,47] we have found that the collection of analytes in a cooled vessel containing a suitable solvent, e.g. DCM for PAHs[23], fitted with a solid phase trap for venting of the supercritical CO_2 and retention of the analytes is suitable for most analytes.

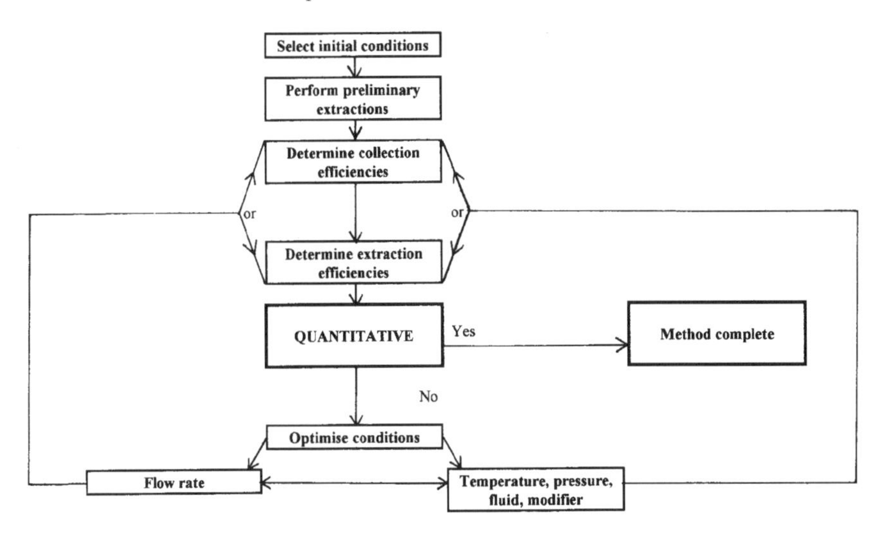

Figure 8.39 Procedure for the development of a quantitative approach to SFE (adapted from Reference 56)

8.5.4 DETERMINE EXTRACTION EFFICIENCY

This involves the determination of the efficiency of the target analytes from real-world samples. Unless you are fortunate the supercritical conditions that have been selected on the basis of spiked samples will not be optimal. At this point it would be ideal to use a certified reference material to establish whether the results obtained are quantitative. However, CRMs are not available for analytes from all matrices, so, it is common to base the results of real-world samples on the recoveries obtained by conventional extraction methods, e.g. Soxhlet extraction. If suitable results are obtained then no additional work is required.

8.5.5 OPTIMISATION OF SFE CONDITIONS

If quantitative results were not obtained it is necessary to investigate the influence of operating variables on the recovery of the target analytes from the real-world samples or CRMs. It is common to investigate the main SFE operating variables: temperature, pressure, modifier addition and flow rate of CO_2. This can be done using a univariate (one at a time) approach or as preferred in our own laboratory a multivariate approach using chemometrics.[23]

REFERENCES

1. C. de la Tour, *Ann. Chim. Phys.*, **21** (1822) 127.
2. J.B. Hannay and J. Hogarth, *Proc. Roy. Soc.* (London), **29** (1879) 324.
3. J.B. Hannay and J. Hogarth, *Proc. Roy. Soc.* (London), **30** (1880) 178.
4. J.B. Hannay, *Proc. Roy. Soc.* (London), **30** (1880) 484.
5. A.W. Francis, *J. Phys. Chem.*, **58** (1954) 1099.
6. K. Zosel, US Patent 3969 196 (1976); French Patent 2, 079, 261 (1971); Belgium Patent 646 641 (1964); German Patent 1, 493, 190 (1964); *Chem. Abstr.*, **63** (1965) 11045b.
7. C.L. Phelps, N.G. Smart and C.M. Wai, *J. Chem. Ed.*, **73** (1996) 1163.
8. M. Novotny, S.R. Springston, P.A. Peaden, J.C. Fjeldsted and M.L. Lee, *Anal. Chem.*, **53** (1981) 407A.
9. J.R. Dean, ed., *Applications of Supercritical Fluids in Industrial Analysis,* Blackie Academic and Professional, Glasgow (1993), p. 49.
10. F.K. Schweighardt and P.M. Mathias, *J. Chromatogr. Sci.*, **31** (1993) 207.
11. J. Via, L.T. Taylor and F.K. Schweighardt, *Anal. Chem.*, **66** (1994) 1459.
12. J.W. King, J.M. Snyder, S.L. Taylor, J.H. Johnson and L.D. Rowe, *J. Chromatogr. Sci.*, **31** (1993) 1.
13. V. Camel, A. Tambute and M. Claude, *J. Chromatogr.*, **642** (1993) 263.
14. V. Janda, K.D. Bartle and A.A. Clifford, *J. Chromatogr.*, **642** (1993) 283.
15. I.J. Barnabas, J.R. Dean and S.P. Owen, *Analyst*, **119** (1994) 2381.
16. T. Greibrokk, *J. Chromatogr.*, **703** (1995) 523.
17. S. Bowadt and S.B. Hawthorne, *J. Chromatogr.*, **703** (1995) 549.
18. I.A. Stuart, J. MacLachlan and A. McNaughton, *Analyst*, **121** (1996) 11R.
19. J.R. Dean, *Analyst*, **121** (1996) 85R.
20. J.J. Langenfeld, S.B. Hawthorne, D.J. Miller and J. Pawliszyn, *Anal. Chem.*, **65** (1993) 338.
21. S.B. Hawthorne and D.J. Miller, *Anal. Chem.*, **66** (1994) 4005.
22. Y. Yang, A. Gharailbeh, S.B. Hawthorne and D.J. Miller, *Anal. Chem.*, **67** (1995) 641.
23. I.J. Barnabas, J.R. Dean, W.R. Tomlinson and S.P. Owen, *Anal. Chem.*, **67** (1995) 2064.
24. I.J. Barnabas, J.R. Dean, I.A. Fowlis and S.P. Owen, *Analyst*, **120** (1995) 1897.
25. J.R. Dean, I.J. Barnabas and I.A. Fowlis, *Anal. Comm.*, **32** (1995) 305.
26. J.R. Dean, *Anal. Comm.*, **33** (1996) 191.
27. S.M. Pyle and M.M. Setty, *Talanta*, **38** (1991) 1125.
28. S.B. Hawthorne, Y. Yang and D.J. Miller, *Anal. Chem.*, **66** (1994) 2912.
29. F. David, M. Verschuere and P. Sandra, *Fresenius J. Anal. Chem.*, **344** (1992) 479.
30. S. Bowadt and B. Johansson, *Anal. Chem.*, **66** (1994) 667.
31. S. Bowadt, B. Johansson, S. Wunderli, M. Zennegg, L.F. de Alencastro and D. Grandjean, *Anal. Chem.*, **67** (1995) 2424.
32. Y. Yang, S. Bowadt, S.B. Hawthorne and D.J. Miller, *Anal. Chem.*, **67** (1995) 4571.
33. H.B. Lee, T.E. Peart and R.L. Hong-You, *J. Chromatogr.*, **605** (1992) 109.
34. S.B. Hawthorne and D.J. Miller, *Anal. Chem.*, **66** (1994) 4005.
35. M.P. Llompart, R.A. Lorenzo and R. Cela, *J. Chromatogr. Sci.*, **34** (1996) 43.
36. M.P. Llompart, R.A. Lorenzo and R. Cela, *J. Chromatogr.*, **723** (1996) 123.
37. J.R. Dean, *J. Chromatogr.*, **754** (1996) 221.
38. J.L. Snyder, R.L. Grob, M.E. McNally and T.S. Oostdyk, *Anal. Chem.*, **64** (1992) 1940.

39. T.R. Steinheimer, R.L. Pfeiffer and K.D. Scoggin, *Anal. Chem.*, **66** (1994) 645.
40. E.G. van der Velde, M. Dietvorst, C.P. Swart, M.R. Ramlal and P.R. Kootstra, *J. Chromatogr.*, **683** (1994) 167.
41. O. Berdeaux, L.F. De Alencastro, D. Grandjean and J. Tarradellas, *Inter. J. Environ. Anal. Chem.*, **56** (1994) 109.
42. J.L. Snyder, R.L. Grob, M.E. McNally and T.S. Oostdyk, *J. Chromatogr. Sci.*, **31** (1993) 183.
43. J. Hedrick and L.T. Taylor, *Anal. Chem.*, **61** (1989) 1986.
44. J. Hedrick and L.T. Taylor, *J. High Res. Chromatogr.*, **13** (1990) 312.
45. J. Hedrick and L.T. Taylor, *J. High Res. Chromatogr.*, **15** (1992) 151.
46. I.J. Barnabas, J.R. Dean, S.M. Hitchen and S.P. Owen, *J. Chromatogr.*, **665** (1994) 307.
47. M. Kane, J.R. Dean, S.M. Hitchen, C.J. Dowle and R.L. Tranter, *Analyst*, **120** (1995) 355.
48. J.L. Ezzell and B.E. Richter, *J. Microcol. Sep.*, **4** (1992) 319.
49. P.H. Tang, J.S. Ho and J.W. Eichelberger, *J. AOAC Int.*, **76** (1993) 72.
50. P.H. Tang and J.S. Ho, *J. High Res. Chromatogr.*, **17** (1994) 509.
51. P.H. Tang, J.S. Ho, J.W. Eichelberger and W.L. Buddle, *J. Chromatogr. Sci.*, **33** (1995) 1.
52. D.C. Messer and L.T. Taylor, J. Chromatogr. Sci., 33 (1995) 290.
53. M. Kane, J.R. Dean, S.M. Hitchen, C.J. Dowle and R.L. Tranter, *Anal. Proc.*, **30** (1993) 399.
54. I.J. Barnabas, J.R. Dean, S.M. Hitchen and S.P. Owen, *Anal. Chim. Acta*, **291** (1994) 261.
55. I.J. Barnabas, J.R. Dean, S.M. Hitchen and S.P. Owen, *J. Chromatogr. Sci.*, **32** (1994) 547.
56. S.B. Hawthorne, D.J. Miller, M.D. Burford, J.J. Langenfeld, S. Eckert-Tilotta and P.K. Louie, *J. Chromatogr.*, **642** (1993) 301.

BIBLIOGRAPHY

T.G. Squires and M.E. Paulaitis, eds, *Supercritical Fluids, ACS Symposium Series 329*, American Chemical Society, Washington, DC, 1987.

B.A. Charpentier and M.R. Sevenants, eds, *Supercritical Fluid Extraction and Chromatography, ACS Symposium Series 366*, American Chemical Society, Washington, DC, 1988.

K.P. Johnston and J.M.L. Penninger, eds, *Supercritical Fluid Science and Technology, ACS Symposium Series 406*, American Chemical Society, Washington, DC, 1989.

M.L. Lee and K.E. Markides, eds, *Analytical Supercritical Fluid Chromatography and Extraction*, Chromatography Conferences, Inc., Provo, UT, 1990.

B. Wenclwaiak, ed., *Analysis with Supercritical Fluids: Extraction and Chromatography*, Springer-Verlag, New York, 1992.

F.V. Bright and M.E.P. McNally, Supercritical Fluid Technology. Theoretical and Applied Approaches to Analytical Chemistry, *ACS Symposium Series 488*, American Chemical Society, Washington, DC, 1992.

S.A. Westwood, ed., *Supercritical Fluid Extraction and its Use in Chromatographic Sample Preparation*, Blackie Academic and Professional, Glasgow, 1992.

K. Jinno, ed., *Hyphenated Techniques in Supercritical Fluid Chromatography and Extraction*, Elsevier Science Publishers, New York, 1992.

J.R. Dean, ed., *Applications of Supercritical Fluids in Industrial Analysis*, Blackie Academic and Professional, Glasgow, 1993.

M. McHugh and V. Krukonis, *Supercritical Fluid Extraction*, Butterworths, Boston, MA, 1994.

M. Saito, Y. Yamauchi and T. Okuyama, eds, *Fractionation by Packed Column SFC and SFE*, VCH Publishers, New York, 1994.

M.D. Luque de Castro, M. Valcarcel and M.T. Tena, *Analytical Supercritical Fluid Extraction*, Springer-Verlag, New York, 1994.

G. Brunner, *Gas Extraction*, Steinkopff Darmstadt-Springer, New York, 1994.

T.A. Berger, *Packed Column SFC*, Royal Society of Chemistry, Cambridge, 1995.

L.T. Taylor, *Supercritical Fluid Extraction*, Wiley-Interscience, New York, 1996.

9

Microwave-Assisted Extraction

Microwave-assisted extraction (MAE) utilises electromagnetic radiation to desorb pollutants from their matrices. The microwave region is considered to exist at wavelengths from 0.3 mm to 1 m and frequencies of 100 GHz to 300 MHz. While the whole of this electromagnetic region is potentially available for use, this is not the case. All microwave ovens (domestic or scientific) operate at 2.45 GHz only.

9.1 THEORETICAL CONSIDERATIONS

The essential components of a microwave system are a microwave generator, wave guide for transmission, resonant cavity and a power supply. At the microwave frequency, electromagnetic energy is conducted from the source to the cavity using a wave guide (or coaxial cable). The microwave generator is a magnetron (a name first coined in 1921 by A.W. Hull). The magnetron is essentially a cylindrical diode, in an axial magnetic field, with a ring of cavities which acts as the anode structure. It is these cavities which become resonant or excited in a way that makes it a source for the oscillations of microwave energy (Figure 9.1). These cavity magnetrons were extensively developed from the 1940s in the USA for terrestrial and satellite communications, i.e. radar and radioastronomy. However, it is the use of microwaves for heating that is important in this book. Historically, the discovery of the magnetron as a heating source was made in 1946 by Percy Spencer who was working for Ratheon in the USA. He noticed that a sweet in his pocket melted when he had been standing close to the magnetron source. He then went on to demonstrate that corn would 'pop' when placed in close proximity to the source. This chance discovery led to the first commercial microwave oven for domestic use appearing in the marketplace in 1967.

The heating effect in microwave cavities is due to dielectric polarisation, i.e. the displacement of opposite charges, of which the most important type is dipolar polarisation. In this situation, the polarisation is achieved by the reorientation of permanent dipoles by the applied electric field. This means that under microwave conditions, a polarised molecule will rotate to align itself with the electric field at a rate of about 10^9 times per second. As the polarisability of a molecule is often represented in terms of the dielectric constant, ε', it is possible to estimate the ability of the microwave to couple to any molecule by considering its ε' values. In the case of microwave-assisted extraction, where samples are heated using organic solvents, values of the dielectric constant for organic solvents are required (Table 9.1).

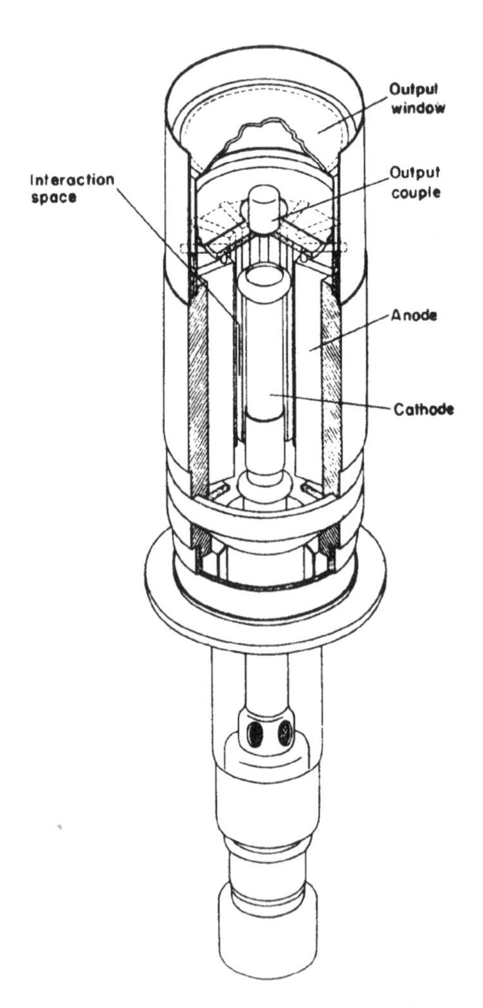

Figure 9.1 Microwave generator: magnetron. From *Encyclopaedic Dictionary of Physics*, Volume 4, Intermediate State to Neutron Resonance Level, J. Thewlis (editor-in-chief), Pergamon Press, Oxford (1961) p. 486

Table 9.1 MAE solvent characteristics[1]

Solvent	Dielectric constant	Boiling Point (°C)	Closed-vessel temperature (°C)[a]
Hexane	1.89	68.7	—
Hexane–acetone	—	52[b]	156
Dichloromethane	8.93	39.8	140
Acetone	20.7	56.2	164
Methanol	32.63	64.7	151
Acetonitrile	37.5	81.6	194

[a] at 175 psig
[b] experimentally determined

9.2 INSTRUMENTATION

The high cost differential between microwave ovens for domestic use and for MAE can often preclude the purchase of a dedicated MAE system. However, for safety reasons (explosions in the presence of organic solvents) it is recommended that only dedicated systems are used. Two types of microwave heating systems are available commercially, the operation of which relates to the sample container arrangement. In the first type, the sample is heated in an open glass vessel fitted with either an air or water condenser while the second approach utilises closed sample vessels constructed in microwave transparent material.

In the open-style system individual sample vessels are heated sequentially. A typical commercial system is the Soxwave from Prolabo Ltd, France, which operates at percentage power increments from 0–100%, corresponding to a maximum of 300 watts. These power increments can be operated in stages and for various time intervals. The sample is introduced into a glass container, along with the appropriate solvent, which is then located within a protective glass sheath. The sample-containing glass vessel, which has the appearance of a large boiling/test tube, is connected to an air or water (preferred option) condenser. This prevents loss of volatile analyte and solvent. At present, this system has no means of measuring the temperature of the solvent. Although an alternative system, the Star system from CEM Ltd, which has been initially developed for inorganic acid digestion, does have an IR sensor for temperature measurement within an open vessel. The system operates, much like Soxhlet apparatus, inasmuch as the organic solvent is seen to be refluxing within the condenser.

A common commercial closed system is the MES-1000 Microwave Solvent Extraction System, as supplied by CEM Corp., USA (Figure 9.2). This system allows up to 12 extraction vessels to be irradiated simultaneously, in 1%

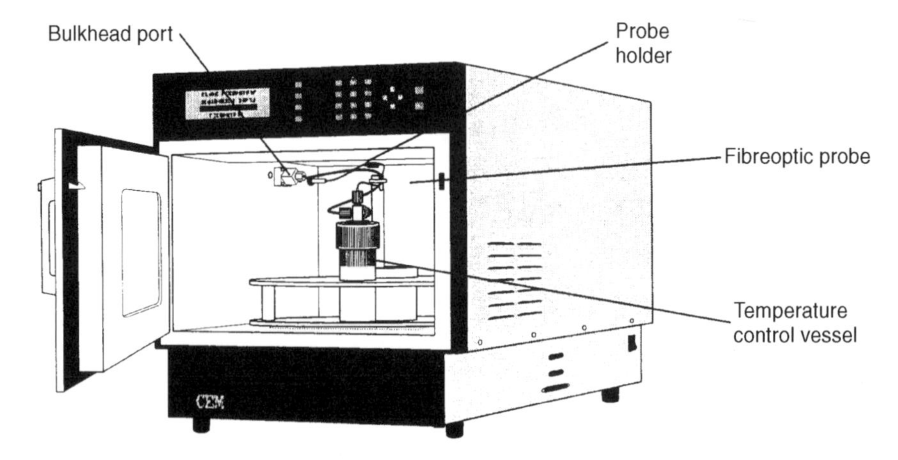

Figure 9.2 Microwave-assisted extraction system. Reproduced by permission of CEM (Microwave Technology) UK Ltd

increments, up to 950 watts of microwave energy at 100% power. The closed system has safety and important experimental features incorporated within its design, most notably, a solvent alarm for the detection of an unexpected release of flammable and toxic organic solvent and an ability to monitor *in situ* both pressure and temperature (within a single extraction vessel only). The pressure is measured using a water manometer that allows readings of up to 200 psi to be made. The temperature probe, a fibre optic with a phosphor sensor,[2] allows extraction temperatures to be selected from 20 to 200 °C in 1 °C increments. The extraction conditions can be varied according to either the percentage power input or by measuring the temperature and pressure within the single extraction vessel. The samples are placed into lined vessels (approximately 100 ml) constructed of polyetherimide (bodies and caps) (Figure 9.3). Inside each vessel is an inner liner and cover, constructed of Teflon perfluoroalkoxy (PFA), with which the sample comes into contact. Each extraction vessel contains a rupture membrane that is designed to fail at elevated pressure (200 psi). Each of the extraction vessels is located in a carousel, which resides within the teflon-lined microwave cavity, which rotates through 180° during microwave operation. In the centre of the carousel and connected to each extraction vessel, is an expansion chamber, which acts to contain escaping vapours in the event of rupture membrane failure. In addition, a solvent detector system, located in the continuously operated system air exhaust, will automatically turn off the magnetron if solvent vapours are detected within the microwave cavity from a leaking extraction vessel. The exhaust fan will continue to operate in the event of a solvent escape.

For suppliers of MAE systems see Appendix A.

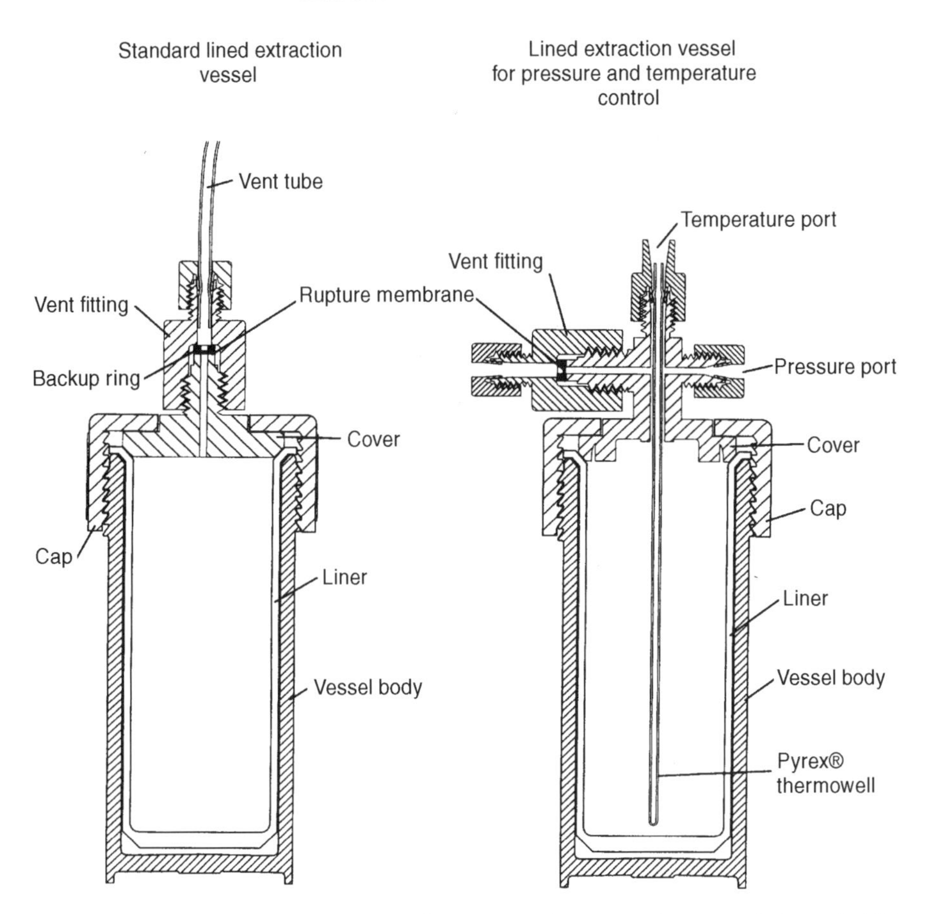

Figure 9.3 Microwave-assisted extraction vessels. Reproduced by permission of CEM (Microwave Technology) UK Ltd

9.3 METHODS OF ANALYSIS: EXTRACTION FROM SOLIDS

The use of a microwave oven for sample preparation was first done in 1975.[3] In this case the microwave oven was used to destroy the organic matter, using combinations of nitric acid with perchloric acid or hydrogen peroxide, of certified reference samples prior to elemental analysis. So while this was the first application of a microwave oven for wet ashing, it is not the topic of this book. The first application of a microwave oven for the microwave-assisted extraction of analytes from matrices using organic solvents did not appear until 1986.[4] In this work, the microwave oven was used to extract analytes from soil, seeds, foods and feeds using methanol or methanol-water for polar compounds and hexane for non-polar compounds. The microwave oven was operated in short 30 s durations and after cooling repeated several times. This approach was compared with the traditional approaches of Soxhlet and shake-flask extraction. In every case, the recoveries obtained by microwave-assisted extraction were comparable with those obtained using the traditional approach. While this paper is of historical importance, its use of a domestic, household microwave oven with organic solvents is not recommended because of fire risk. Modern commercial systems for microwave-assisted extraction have safety features (see experimental) which minimise the risks involved in heating organic solvents.

This section will review the application of microwave-assisted extraction for analytes of environmental origin with a view to recommending the most appropriate operating conditions. (Note: Unless otherwise stated, MAE is operated under pressurised conditions.)

9.3.1 POLYCYCLIC AROMATIC HYDROCARBONS (PAHS)

Polycyclic aromatic hydrocarbons are a class of compounds that are found on soils as a consequence of previous heavy industry.

As part of an ongoing programme, by the US Environmental Protection Agency, addressing sample preparation techniques to prevent or minimise pollution in analytical laboratories MAE was evaluated for the extraction of 17 PAHs and a few base/neutral/acidic compounds from spiked soil samples and certified reference materials.[5] A systematic approach to investigate the effect of two MAE operating variables was performed. The variables considered were: temperature (80, 115 and 145 °C); and extraction time, (5, 10 and 20 min) using a 1 + 1 solvent mixture of hexane : acetone (30 ml). The results were compared with room temperature extraction, in which the solvent mixture was allowed to stand with the soil for the same length of time as MAE including cooling time. Six CRMs were evaluated, four sediments (HS-3, HS-4 and HS-5 available from the NRCC and a NIST SRM 1941) and two certified soils (SRS 103-100 and ERA, lot no. 321). Whereas the average recovery at room temperature was ~52%, the MAE recoveries were 70% at 80 °C, 75% at 115 °C and 75% at 145 °C. Since the recoveries at 115 and 145 °C were almost identical, a temperature of 115 °C was used for further work. An extraction time of 5 min was found to be adequate; no statistical difference at the 95% confidence interval was noted when the extraction time was increased up to 20 min. Further work from the same group[6] using a 1 + 1 hexane : acetone solvent mixture, at a temperature of 115 °C and an extraction time of 10 min was done for 187 compounds including PAHs from spiked soils that were either extracted immediately or aged for 24 h, 14 days or 21 days. It was found that the recoveries for the aged samples usually decreased and that there was more spread in recoveries with ageing.

Barnabas *et al.*[7] compared a commercial MAE system with a 6-hour Soxhlet extraction for the removal of PAHs from highly contaminated soil. Initial work evaluated the effect of MAE solvent. Previous work published in the literature had identified a combination of hexane and acetone to be an effective solvent mixture for the extraction of a range of pollutants from soils.[5,6] In this work, the effect of varying the hexane-acetone ratio was evaluated. The combinations were varied from 80 : 20 hexane : acetone through to 0 : 100 hexane : acetone in steps of 10. It was found that as the solvent mixture became more polar i.e. a hexane : acetone ratio of 10 : 90 or higher (in favour of acetone) it was possible to recover higher concentrations of PAHs from soil. Subsequent work used 100% acetone only. To evaluate the three main operating conditions, a chemometrics approach, based on a central composite design was used. The operating conditions varied were: temperature; extraction time; and solvent volume between upper and lower limits of, 120 and 40 °C; 20 and 5 min; and, 30 and 50 ml, respectively. In total 20 experiments were required to evaluate the central composite design. The results, based on total recovery of 16 PAHs, indicated that within the limits of the operating variables considered no dependence on operating variables was evident. Using a temperature of 120 °C, extraction time of 20 min and 40 ml of 100% acetone the repeatability obtained was excellent for each individual PAH with an average recovery, based on 16 PAHs, of 422.9 $mg\,kg^{-1}$ with an RSD of 2.4%. An interlaboratory test soil, CONTEST, as supplied by the Laboratory of the Government Chemist was extracted using both Soxhlet extraction with DCM as the solvent and MAE using either acetone or DCM as the solvent. In all cases the results were in agreement with each other.

A further approach to optimise a commercial MAE system for the extraction of PAHs from two marine sediment certified reference materials (HS-4 and HS-6, obtained from the NRCC, Canada) was reported.[8] In this approach a mixed-level orthogonal design was used. The four variables optimised were: types of extraction solvent, extraction temperature, extraction time and volume of extraction solvent. The extraction solvents considered were: DCM, acetone–hexane (1 + 1), acetone–petroleum ether (1 + 1) and methanol–toluene (9 + 1). The levels of the other three variables considered were as follows: temperature, 115 and 135 °C; extraction time, 5 and 15 min, and solvent volume, 30 and 45 ml. The results indicated that statistical significance ($p < 0.1$) was evident for both the individual parameters of extraction solvent and temperature and their two-variable interaction term. The optimum MAE conditions for the extraction of 16 PAHs from marine sediment were determined to be: 30 ml of hexane : acetone (1 + 1); temperature, 115 °C; and, an extraction time of 5 min. As both HPLC with fluorescence detection and GC-MSD were used for analysis, it was necessary to extract samples in DCM and then solvent exchange to methanol for the HPLC analysis only. This optimised approach was compared with a 16-hour Soxhlet extraction using 300 ml of DCM as the solvent. The results are shown in Figure 9.4 for HS-4 and Figure 9.5 for HS-6. Good agreement was obtained between the two extraction approaches. The MAE

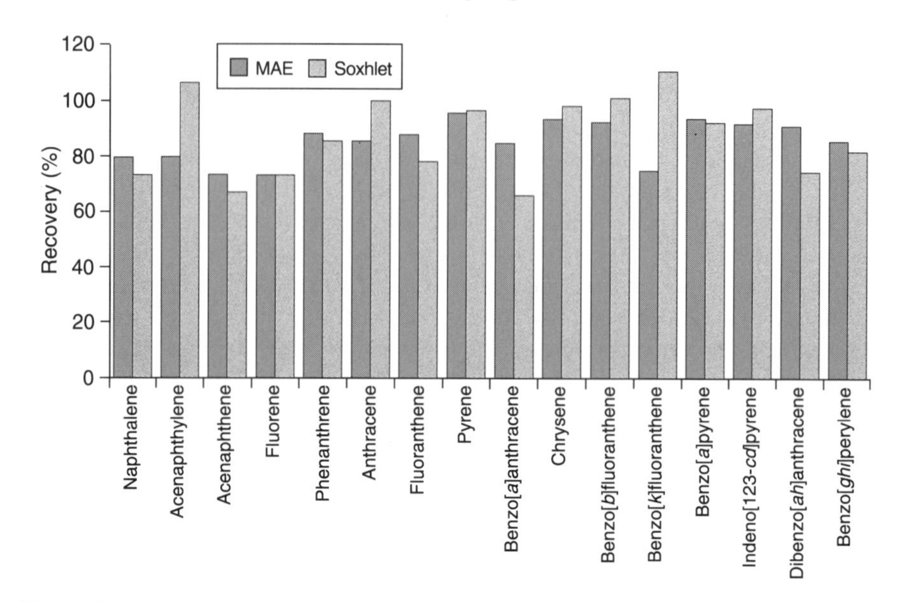

Figure 9.4 Extraction of polycyclic aromatic hydrocarbons from marine sediment (HS-4): comparison of Soxhlet and MAE. From Chee *et al.*, *Journal of Chromatography*, **723** (1996) 259

method was then applied to four Singapore coastal sediment samples. Two of the 16 PAHs were not detected in the samples (naphthalene and benzo[*a*]anthracene) while the results for the other 14 individual PAHs ranged from 0.03 to 0.35 mg kg^{-1}. The speed of extraction (12 samples in one run in less than 30 min) and the lower solvent consumption (30 ml per sample) compare favourably with Soxhlet extraction.

9.3.2 PESTICIDES

Onuska and Terry[9] slurry spiked air-dried sediment with a range of organochlorine pesticides and then left the samples to age for at least one month prior to MAE extraction. This paper reports the first results from Environment Canada who have patented the process,[10] using the terminology, Microwave Assisted Process (MAP^{TM}). The results highlighted the dependence on organic extraction solvent; iso-octane : acetonitrile 1 : 1 being preferred to the individual solvents; sediment sample moisture, a mimimum water content is necessary to perform MAE with the best results obtained with 15% water level; and, extraction time, at least 3 min required. The results, based on $n = 5$, of the extraction of 15 OCPs from a sediment sample spiked in the range 50–250 μg kg^{-1} were good, the minimum recovery was

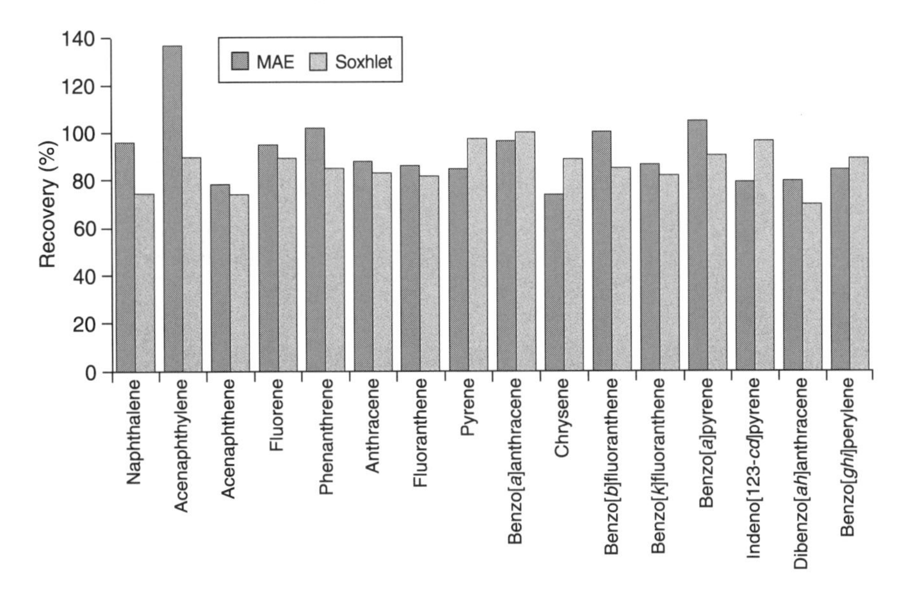

Figure 9.5 Extraction of polycyclcic aromatic hydrocarbons from marine sediment (HS-6): comparison of Soxhlet and MAE. From Chee *et al.*, *Journal of Chromatography*, **723** (1996) 259

74% with an RSD of 3% for p,p'-DDE and the best recovery was 95.3% for methoxychlor with an RSD of 3.9%.

The recovery of 20 OCPs from spiked soils using MAE was reported by Lopez-Avila *et al.*[11] Using the following MAE conditions: temperature, 115 °C; extraction time, 10 min at 100% power; and 30 ml of a 1 : 1 acetone : hexane solvent mixture, the recoveries of the OCPs spiked at the 50 ng g^{-1} level were assessed from clay soil, topsoil, sand, organic compost and topsoil with 5% humic acid. The results showed that the sand matrix produced the cleanest extracts, the highest recoveries (mean recovery of the 20 OCPs was 83.4% with an RSD of 10.6%) and the best precision (typically 2–3%, based on three determinations). Recoveries from the other matrices were only slighly different (clay soil, mean recovery 85.1% with an RSD of 20%, $n = 19$; and organic compost, mean recovery 88.8% with an RSD of 30.0%, $n = 16$) with poorer recoveries being obtained from the top soil with 5% humic acid added (mean recovery 71.9% with an RSD of 24.9%, $n = 20$) and the topsoil only (mean recovery 72.9% with an RSD of 20.4%, $n = 20$). The poorer precision in all cases should be noted.

The MAE of OCPs from spiked soil and a certified reference material was reported by Fish and Revesz.[12] The spiking procedure was as follows: 5 g of soil was added directly to the extraction vessel, then the spike was added and stirred with a glass rod. The vessel was then sealed and allowed to equilibriate for 1 h. The

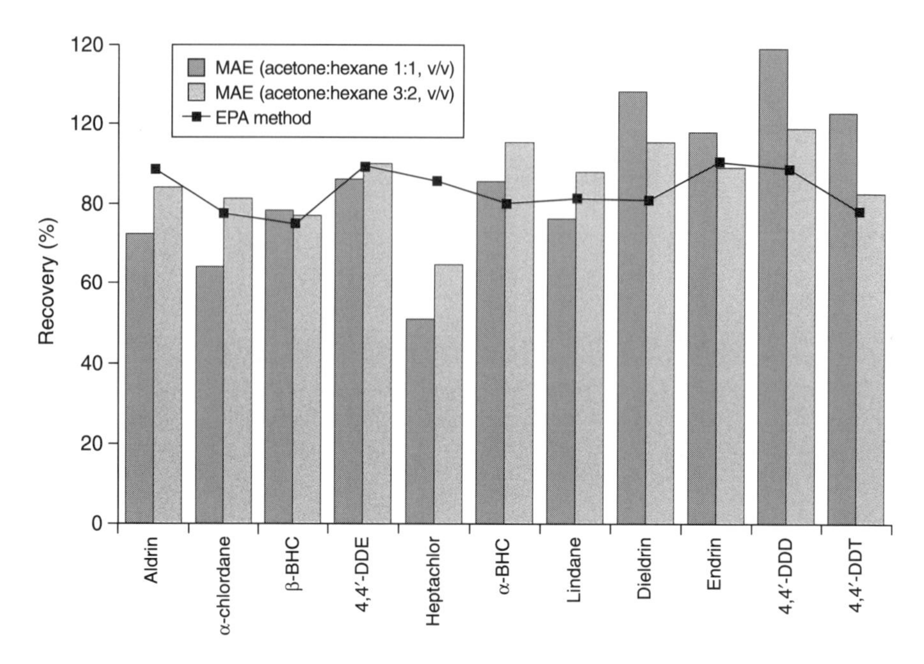

Figure 9.6 Microwave-assisted extraction of organochlorine pesticides from spiked soil samples: comparison with EPA laboratory results. From Fish and Revesz, *LC–GC*, **14** (1996) 230

MAE was done using 30 ml of a 1 : 1 acetone : hexane mixture with heating for 15 min at various temperature settings from 90, 110 and 120 °C. As the recoveries increased in accordance with temperature increases, the maximum temperature evaluated (120 °C) was selected. The procedure was then applied to a CRM using 3 g of the soil with MAE conditions as previously determined. In addition, the extraction solvent ratio was altered to 3 : 2 acetone : hexane to match the azeotrope's vapour phase ratio. The results, compared with values obtained from 10 EPA laboratories, are shown in Figure 9.6. The best results were obtained using 50 ml of the 3 : 2 acetone : hexane solvent mixture.

The MAE of OCPs from seal blubber and pork fat with n-hexane as the solvent has been described.[13] It is generally accepted that hexane is not the most suitable solvent for MAE, due to its poor dielectric constant. However, to overcome this problem a microwave transformer (Weflon®, an inert material based on carbon-containing teflon) was used to transfer energy to the sample. In order to prevent solvent and analyte leakage, the MAE system was operated in extraction cycles of short time durations (30 s) at full power followed by 5 min cooling. As no certified reference samples were available for this work, the blubber of a grey seal stranded on Rugen, Baltic Sea, Germany was used. This seal blubber was initially melted, subjected to matrix clean-up with deactivated silica and analysed. This provided a reference point for the MAE work which was done on the naturally contaminated blubber. For all organochlorine compounds studied the best recoveries were obtained using seven MAE extraction cycles, average recoveries for the seven compounds was $96.9 \pm 0.5\%$. Spiking experiments on pork fat of the organochlorine compounds gave excellent recoveries ($88.5\% \pm 1.3\%$ to $98.6\% \pm 2.3\%$) based on seven MAE cycles and $n = 5$.

9.3.3 HERBICIDES

The extraction of atrazine and its principal degradates from agricultural soil was carried out by Steinheimer.[14] In this case the herbicides were extracted using a microwave oven with organic-free water and 0.35 M HCl as the extraction solvent. The author noticed that cleaner extracts and simpler HPLC chromatograms were observed using this approach compared to the more commonly used methanol-water and acetonitrile-water mixtures. However, the mean recoveries for atrazine were only in the range 55–65%. The typical standard deviation, based on 25 replicates, was approximately 30–40%.

The MAE of a significant new class of low-use-rate, reduced-environmental-risk herbicides for the protection of a wide variety of agricultural commodities, the imidazolinones, have been extracted from a variety of soil types fortified at the 1–50 ppb range.[15] The aim of this work was to extract imazethapyr, the most widely used imidazolinone, from soil and be able to determine its concentration at the 1 ppb level. It was found that the following operating conditions: temperature, 125 °C; extraction time, 3 min; and, an extraction solvent of 0.1 M NH_4OAc/NH_4OH at pH 10, gave acceptable recoveries, i.e. an average recovery of $92 \pm 13\%$ based on soil spikes in the range 1 to 50 ppb ($n = 12$).

9.3.4 PHENOLS

The continued interest by USEPA in new sample preparation techniques has focused on a wide range of compounds including phenols.[11] In this work 16 phenolic compounds were MAE from clay soil, topsoil, sand, organic compost and topsoil with 5% humic acid using the following conditions: temperature, 115 °C; extraction time, 10 min at 100% power; and 30 ml of a 1 : 1 acetone : hexane solvent mixture. The recoveries of the phenolic compounds, spiked at the 50 $\mu g\, g^{-1}$ level, were assessed. The results indicated that 70% of the total samples analysed gave recoveries above 70%. In addition, 2,4-dinitrophenol and 4,6-dinitro-2-methylphenol appear to have degraded or are irreversibly adsorbed; no such effect was noted for the extraction from sand. The precision of the data, in most cases, is excellent with RSD values of $< 10\%$ in 93% of samples extracted and analysed.

A chemometric approach, based on a central composite design, has been applied for the MAE of phenol and methylphenol isomers from spiked soils that have been aged for 25 days.[16] The variables considered and their limits were as follows: temperature, 70–130 °C; proportion of acetone, 20–80%; and solvent volume, 15–50 ml, at a fixed extraction time of 10 min and for 5 g of sample. Temperature was found to be significant for all four phenols studied while the proportion of acetone to hexane and the solvent volume were found to be significant in some cases. It was concluded that 10 ml of acetone : hexane (80 : 20) at 130 °C was the most appropriate, producing good recoveries.

9.3.5 POLYCHLORINATED BIPHENYLS (PCBS)

Polychlorinated biphenyls have been extracted from a range of certified reference materials (marine sediment and soils) by MAE using the following conditions: solvent, 30 ml of hexane : acetone (1 + 1); temperature, 115 °C; and, extraction time, 10 min.[6] The results for a range of Aroclors (1254, 1260, 1016, 1248) were in agreement with certificate values. The same group has compared the method of detection of the PCBs after MAE.[17] The detection methods compared were GC with ECD or ELISA. The rapidity of MAE and ELISA allows a batch of 10 samples to be prepared and analysed in approximately 1 hour. In addition, cost saving in terms of solvent usage are evident compared to Soxhlet extraction.

9.3.6 PHTHALATE ESTERS

Phthalate esters have been extracted from marine sediment and soil as part of an ongoing ASEAN-Canada Cooperative Programme on marine science (ASEAN refers to a group of South-East Asian countries that includes Singapore, Malaysia, Thailand, Brunei, Indonesia and the Philippines).[18] In this work the effects of extraction solvent, solvent volume, temperature and extraction time were evaluated for the removal of six phthalate esters from marine sediment. The six phthalates investigated were dimethyl phthalate, diethyl phthalate, diallyl phthalate, dibutyl phthalate, benzyl n-butyl phthalate and di(2-ethylhexyl)phthalate. The four operating parameters were evaluated from a spiked sample matrix as follows: extraction solvent, dichloromethane, acetone/hexane and acetone/petroleum ether; solvent volume, 25, 30 and 35 ml; temperature, 80, 115 and 145 °C; and extraction time, 5, 10 and 15 min. It was concluded that in the case of extraction solvent, no statistical difference was observed between a 1 : 1 acetone : hexane mixture and DCM; it was preferable to use the solvent mixture for two reasons. The first reason is that no solvent exchange is required prior to analysis when using the hexane:acetone mixture and secondly that DCM is not a suitable solvent for GC with ECD. The effects of temperature and extraction were related. An increase from the lowest operating condition i.e. temperature 80 °C and extraction time 5 min, resulted in a significant increase in recovery. However, no significant increase in recovery was noted when the operating conditions were raised from their mid-point to the highest condition of operation. However, solvent volume

30 ml was considered to be the most appropriate as it offered the highest recoveries in the majority of cases (five out of six). The optimum conditions were therefore identified as: 30 ml of a 1 : 1 solvent mixture (hexane : acetone), at a temperature of 115 °C for 10 min. This method of extraction was then applied to marine sediment located in the Tuas/Jurong industrial area of Singapore. The results obtained indicated that dibutyl phthalate and di(2-ethylhexyl) phthalate were present in the ranges 0.68–1.60 mg kg^{-1} and 0.16–2.79 mg kg^{-1}, respectively.

9.3.7 ORGANOMETALLICS

Atmospheric MAE has been applied to the extraction of organotin compounds from sediment samples.[19] Two classes of tin compounds were investigated, butyl and phenyl substituted i.e. monobutyltin, dibutyltin and tributyltin; and monophenyltin, diphenyltin and triphenyltin. The effects of microwave power (20, 60, 100 and 160 watt) and extraction solvent (isooctane, methanol, water and artificial seawater) were investigated for each of the compounds individually spiked on a clean (< 5 ng g^{-1} as tin) sediment sample. It was concluded that the use of an organic solvent, i.e. methanol, was advisable and that microwave exposure times should not exceed 10 min and 100 W power. It was observed that conditions outside these led to rupture of the tin-carbon bond and the formation of inorganic tin. The system was then investigated using a certified reference material (PACS-1). It was found that the addition of 10 ml of a 0.5 M acetic acid in methanol at an optimum power of 60 W for 3 min was necessary. The procedure was then applied to a further certified reference material (CRM 462). The results, shown in Figure 9.7 for CRM 462 and Figure 9.8 for PACS-1, show the effectiveness of this approach. However,

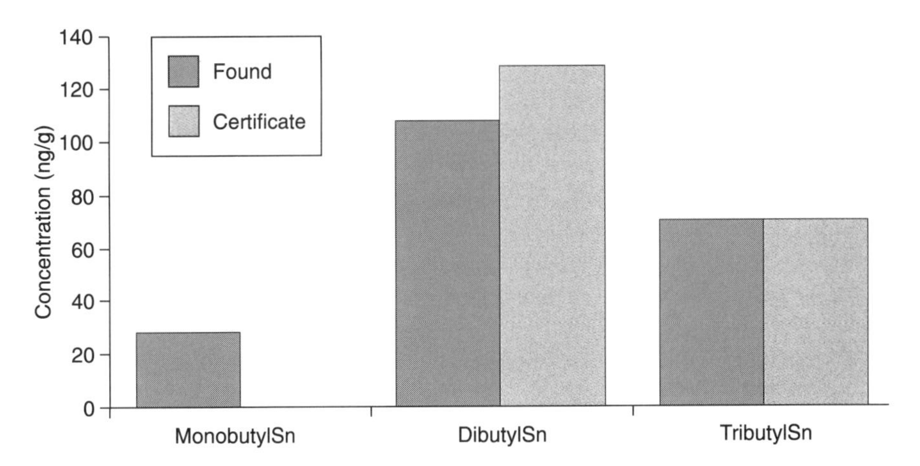

Figure 9.7 Atmospheric MAE of organotin compounds from CRM 462. From Donard *et al.*, *Analytical Chemistry*, **67** (1995) 4250

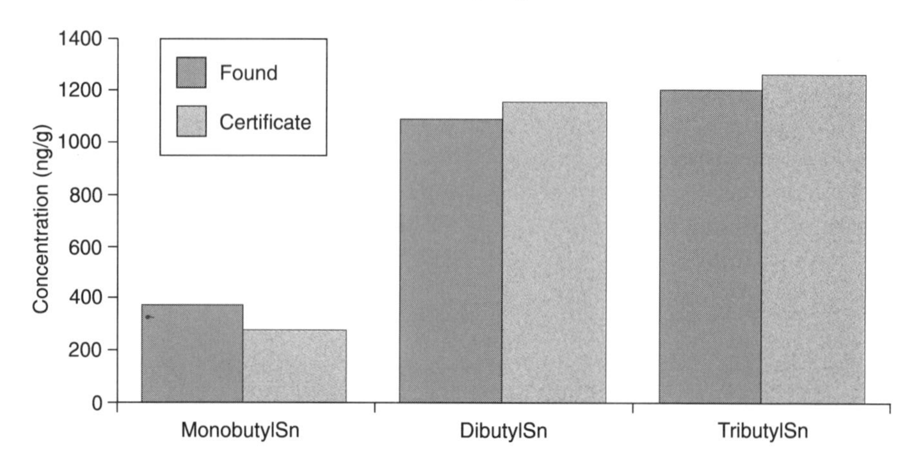

Figure 9.8 Atmospheric MAE of organotin compounds from PACS-1. From Donard *et al.*, *Analytical Chemistry*, **67** (1995) 4250

care should be taken to optimise the microwave technique on real samples and not spiked samples, as highlighted by this work.

Atmospheric MAE has also been applied by Szpunar *et al.*[20] for the leaching of mono-, di-, and tributyltin from sediments and biomaterials. In this procedure 50% acetic acid was used for the sediment samples and 25% tetramethyl ammonium hydroxide for the biomaterials using a low-power microwave field (60 W) within 1–5 min. In each case the butyltins were then derivatised with sodium tetraethylborate and extracted into hexane prior to analysis by large volume injection GC coupled to a quartz furnace atomic absorption spectrometer. Two sediment CRMs (PACS-1 and BCR 462) were analysed and the results obtained agreed with their certificate values for dibutyltin and tributyltin, the exception was monobutyltin. For this tin species the results obtained in the PACS-1 CRM were higher ($1.06 \pm 0.1\ \mu g\ g^{-1}$) compared to the certificate value ($0.28 \pm 0.17\ \mu g\ g^{-1}$). The authors believe that this particular species has not been fully recovered from the sediment by the other methods reported. They present evidence that indicates the wide variability in monobutyltin content in both PACS-1 and BCR 462 as reported in the literature. A further CRM was also analysed (NIES11 fish tissue) which has a certified value of $1.3 \pm 0.1\ \mu g\ g^{-1}$ for tributyltin which compared favourably with the result reported in this work ($1.2 \pm 0.1\ \mu g\ g^{-1}$).

Pressurised MAE for the leaching of methylmercury from sediment samples has been reported by Vazquez *et al.*[21] Two- and three-level factorial designs were used to optimise the parameters for MAE. The final optimised conditions were: temperature, 120 °C; 6 M HCl, 400 μl; toluene, 10 ml; and extraction time, 10 min. The procedure was then applied to two sediment samples, BCR S-19 and a candidate CRM 580. The results showed good agreement between the pressurised

MAE approach, a conventional acid leaching/manual extraction procedure and the intercomparison value.

9.4 METHODS OF ANALYSIS: EXTRACTION FROM WATER

A novel approach for the microwave oven has been its use to extract various classes of compounds from water samples. In a unique approach[22] an aqueous sample was placed in a screw-capped container located in the microwave oven (Figure 9.9). Through the cap of the container was located a helium purge line and an exhaust line. The exhaust line led to a trap which was cooled in an ice bath. The procedure of operation was as follows: The 1-litre sample, to which was added 20 g of salt, was purged with helium for 20 min while operating the microwave oven at full power for 7 min (assessed on the basis of safety constraints). The collected extract was then subjected to further clean-up prior to analysis for chlorinated benzenes. This approach was compared with the traditional approach of liquid–liquid extraction and found to compare favourably at the $ng\,l^{-1}$ level (Figure 9.10). The same group[23] also applied the approach to the extraction of PCBs from 500 ml of water. Recoveries of the spiked congeners, at two levels 85.5 and $175\,pg\,l^{-1}$, in organic-free water were in the range 68–85%. The technique was applied to Hamilton Harbour water and the results compared with liquid–liquid extraction (Figure 9.11) where it can be seen that excellent agreement is obtainable.

In what seems to be at first glance a rather odd approach, the microwave has been used to elute pollutants from a solid phase extraction (SPE) membrane.[24]

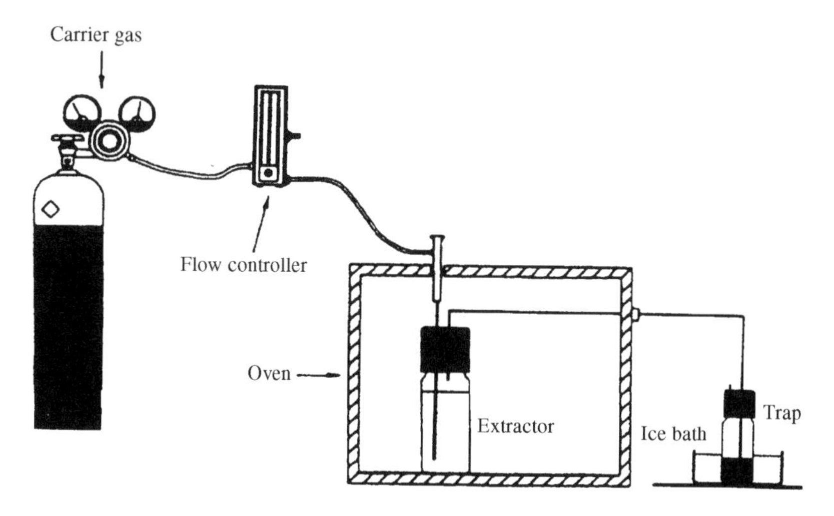

Figure 9.9 Schematic diagram of the dynamic microwave extraction system. From Onuska and Terry, *Journal of Microcolumn Separations*, **7** (1995) 319

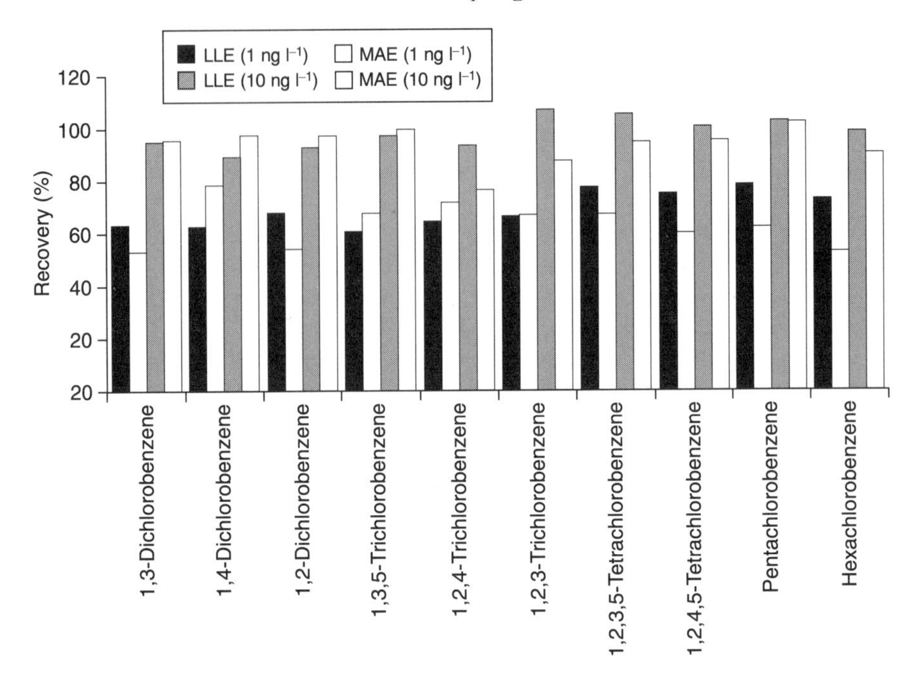

Figure 9.10 Comparison of MAE with LLE for chlorinated benzenes from spiked water samples. From Onuska and Terry, *Journal of Microcolumn Separations*, **7** (1995) 319

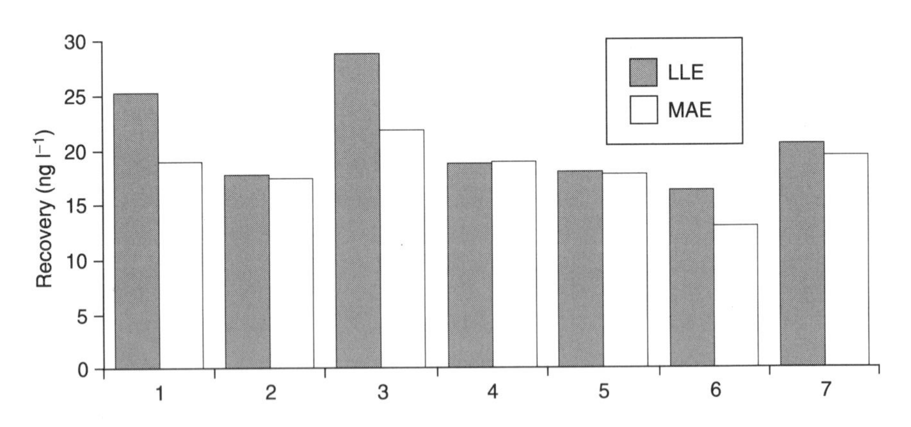

Figure 9.11 PCB 209 from Hamilton Harbour water[23]

Pollutants are retained on a C18 membrane disk, rolled up and transferred into the closed PTFE-lined vessels, prior to elution with organic solvent using the microwave oven. This approach was applied to the extraction of OCPs, PCBs, PAHs, phthalate esters, OPPs, fungicides, herbicides and insecticides. Two elution solvents (acetone and DCM) were evaluated as were the following microwave parameters: temperature (80, 100 and 120 °C) and extraction time (1, 3, 5 and 10 min) at 50% power. After optimisation the following were selected: solvent, acetone; temperature, 100 °C; and extraction time, 7 min. The results compare the recoveries obtained for: PAHs and phthalates (spiked at 2 $\mu g\,l^{-1}$); OCPs and PCBs (spiked at levels between 0.1 and 0.2 $\mu g\,l^{-1}$ for OCPs and 0.5 $\mu g\,l^{-1}$ for PCBs); and, OPPs, fungicides, insecticides and herbicides (spiked at levels between 1 and 2 $\mu g\,l^{-1}$) in either sea water (Figure 9.12) or reagent water (Figure 9.13). It can be seen that the proposed SPE-MAE approach compares favourably with the traditional approach of liquid–liquid extraction.

9.5 GAS-PHASE MICROWAVE-ASSISTED EXTRACTION

The utilisation of a microwave oven to liberate volatile organic compounds (VOCs) from water samples has been described by Pare *et al.*[25] In this process, patented by Environment Canada as microwave-assisted process (MAP™), the sample is subjected to heat, via a microwave oven, and VOCs are vaporised from the water matrix into the headspace above. At this point a conventional headspace sampler is used to introduce the VOCs into a GC with FID. The results reported[25] are compared with a conventional 30-min static headspace sampling apparatus. It was found that the microwave approach gave higher detector responses, with better precision and in a shorter time scale than the conventional approach.

9.6 COMPARISON WITH OTHER EXTRACTION TECHNIQUES

Microwave-assisted extraction has been compared with Soxhlet, sonication and SFE for the extraction of 94 compounds as listed in USEPA method 8250.[26] Freshly spiked soil samples and two standard reference materials were extracted using MAE (conditions: sample, 10 g; 30 ml of hexane : acetone, 1 : 1; temperature, 115 °C; and an extraction time of 10 min), Soxhlet extraction (conditions: sample, 10 g; 300 ml of hexane : acetone, 1 : 1; and an extraction time of 18 h); sonication (conditions: sample, 30 g; 100 ml of methylene chloride : acetone, 1 : 1 repeated three times; and an extraction time of 3 min); and SFE (conditions: sample, 5 g; 10% methanol-modified supercritical CO_2; pressure, 450 atm; temperature, 100 °C; and an extraction time of 60 min). For the results reported for the 94 compounds, 51 compounds gave MAE recoveries of $>80\%$; 33, 50–79%; 8, 20–49%; and 2,

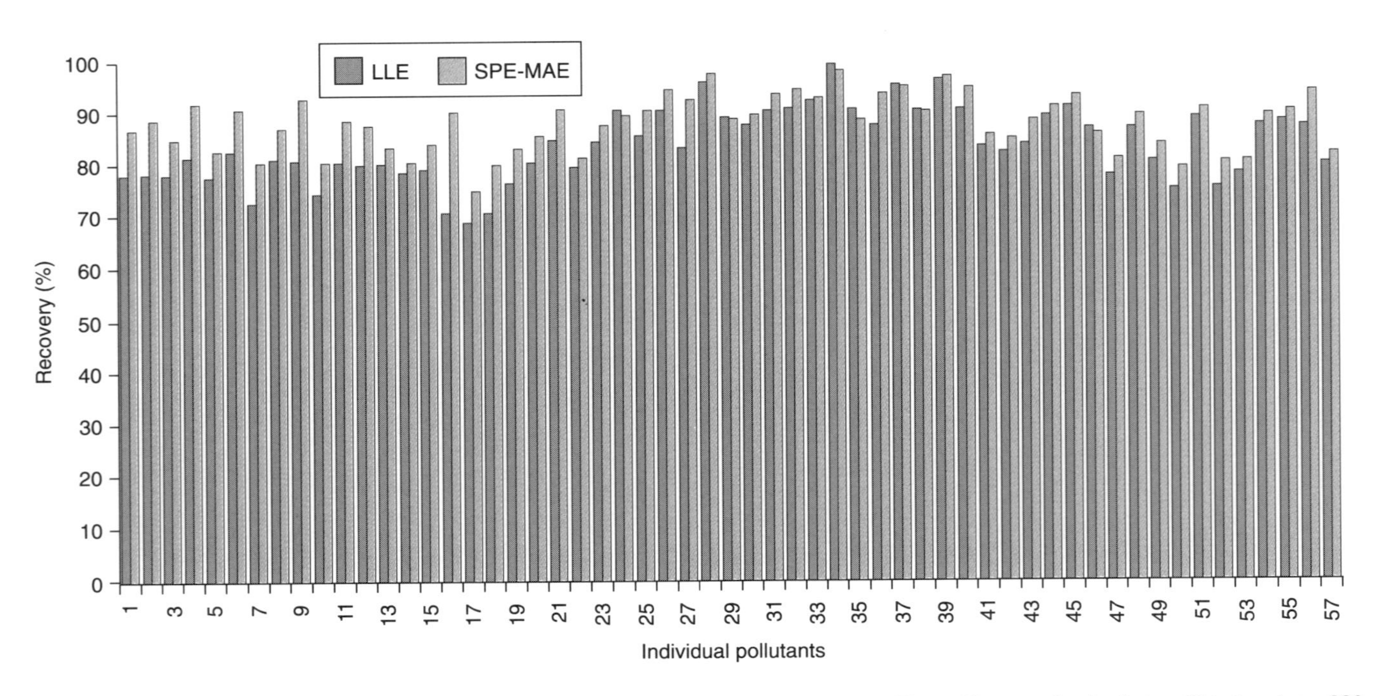

Figure 9.12 Comparison of SPE-MAE with LLE: recoveries obtained from sea water. From Chee *et al.*, *Analytica Chimica Acta*, **330** (1996) 217

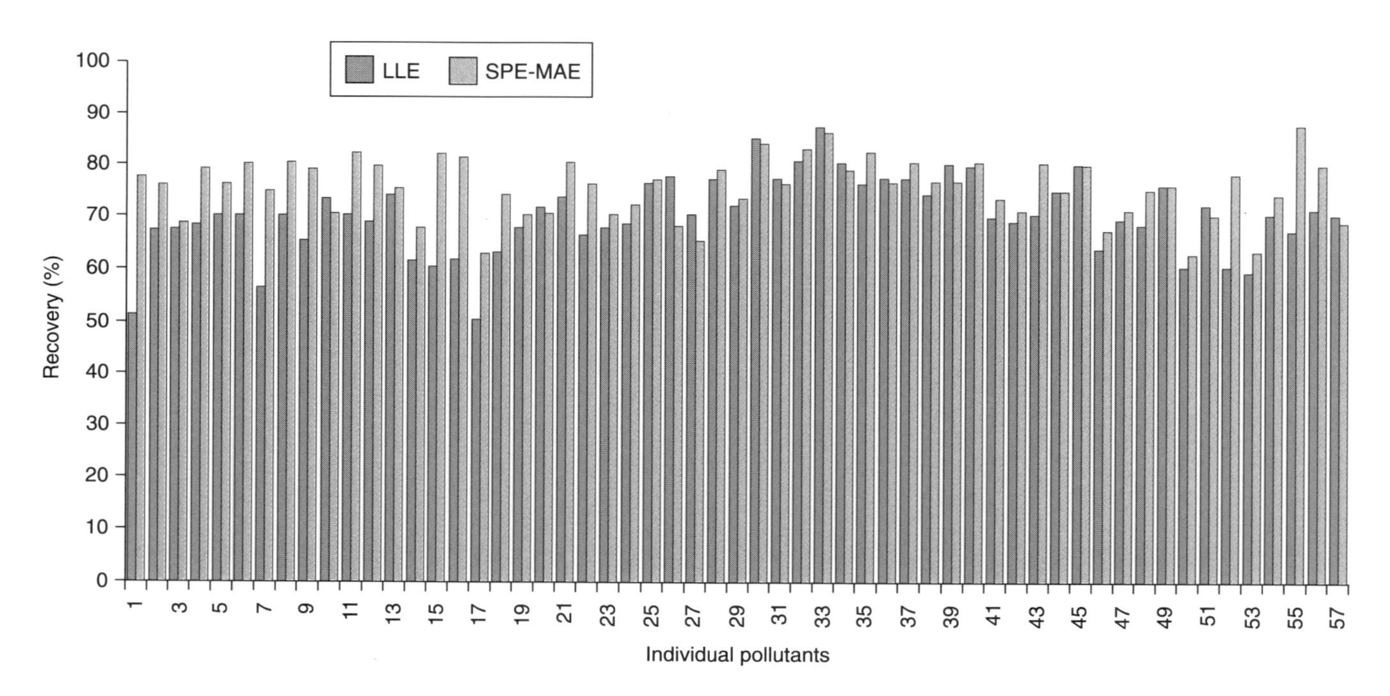

Figure 9.13 Comparison of SPE-MAE with LLE: recoveries obtained from reagent water. From Chee *et al.*, *Analytica Chimica Acta*, **330** (1996) 217

$< 19\%$. Soxhlet extraction gave similar results: 50 recoveries were $> 80\%$; 32, 50–79%; 8, 20–49%; and 4, $< 19\%$. Sonication recoveries were slightly higher: 63 values were $> 80\%$; 25, 50–79%; 4, 20–49%; and 2, $< 19\%$. SFE recoveries were the lowest: 37 values were $> 80\%$; 37, 50–79%; 12, 20–49%; and 8, $< 19\%$. The best precision was obtained using MAE; RSDs $\leqslant 10\%$ for 90 out of 94 compounds studied. Soxhlet extraction gave the worst precision: only 52 out of 94 samples gave RSDs $\leqslant 10\%$.

9.6.1 POLYCYCLIC AROMATIC HYDROCARBONS (PAHS)

Dean *et al*[27] have compared MAE with Soxhlet and SFE for the extraction of PAHs from contaminated soil. The authors reported that MAE with a polar organic solvent, i.e. acetone, gave excellent recoveries (total recovery of 16 individual PAHs, 422.9 mg kg^{-1}) whereas MAE with a less polar solvent, i.e. DCM gave poorer recoveries (total recovery of 16 individual PAHs, 279.8 mg kg^{-1}). This lower recovery was in agreement with the recovery obtained after a 6-hour Soxhlet extraction using DCM (total recovery of 16 individual PAHs, 297.4 mg kg^{-1}). The highest recovery compared favourably with the results obtained using supercritical CO_2 with 20% methanol (total recovery of 16 individual PAHs, 458.0 mg kg^{-1}). The use of a polar solvent is recommended for MAE at a temperature of 120 °C and an extraction time of 20 min.

9.6.2 ORGANOCHLORINE PESTICIDES

Organochlorine pesticides, spiked on soil at the 50 $\mu g\ kg^{-1}$ level for the sonication and Soxhlet extractions and 20 $\mu g\ kg^{-1}$ for MAE (except hexachlorobenzene and hexachlorocyclopentadiene which were spiked at 200 and 100 $\mu g\ kg^{-1}$, respectively), were compared with MAE.[6] The results shown in Figure 9.14 indicate that the highest recoveries were obtained by MAE, in most cases.

9.6.3 PHENOLS

Six phenols were extracted from an ERA soil (lot no. 330) by MAE, Soxhlet and sonication.[11] Both Soxhlet and sonication were done in accordance with EPA methods, 3540 and 3550, respectively. The highest recoveries were obtained by MAE (71%) compared to Soxhlet (52%) and sonication (57%). The precision of MAE was also far superior to both other extraction techniques.

A comparison between MAE and sonication was done by Llompart *et al.*[16] for the extraction of phenol and methylphenol isomers from spiked soil samples after optimisation of the MAE conditions using a central composite design. The results (Figure 9.15) clearly indicate the superior methodology of MAE compared to sonication.

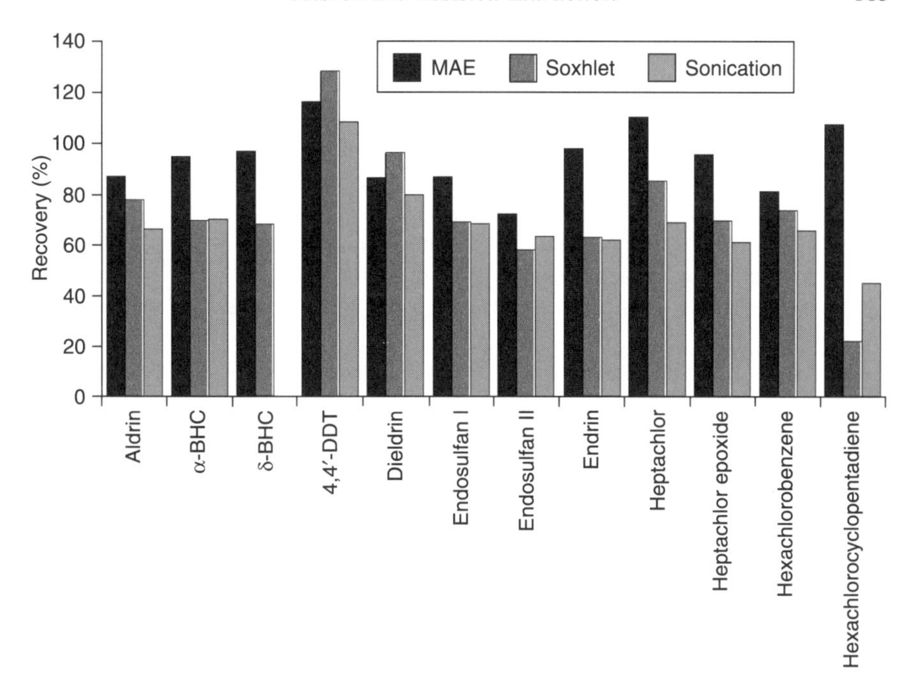

Figure 9.14 Extraction of organochlorine pesticides from spiked soil: comparison between MAE, Soxhlet and sonication. From Lopez-Avila *et al.*, *Analytical Chemistry*, **67** (1995) 2096

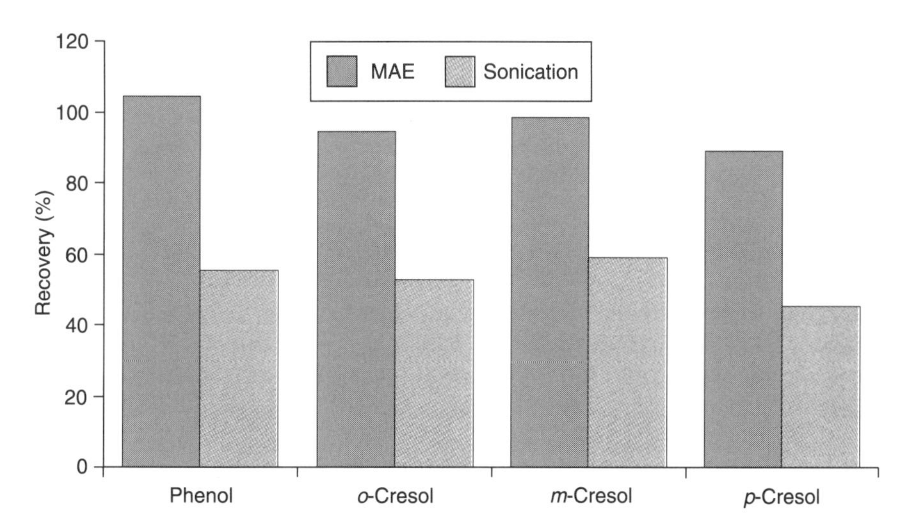

Figure 9.15 Extraction of phenol and methylphenol isomers from spiked soil: comparison between MAE and sonication. From Llompart *et al.*, *Analyst*, **122** (1997) 133

9.6.4 PHTHALATE ESTERS

The extraction of six phthalate esters from marine sediment was evaluated by comparing an optimised MAE approach with sonication and Soxhlet extraction.[18] The MAE was performed according to the following conditions: 30 ml of a 1 : 1 solvent mixture (hexane : acetone), at a temperature of 115 °C for 10 min. Soxhlet was done using 300 ml of DCM and extracting for 16 h while sonication was done using 50 ml of DCM subjected to sonication in an ultrasonic bath for 15 min and repeated a further two times. Using spiked marine sediment the mean recoveries of the phthalate esters were 70.9–91.0%, 65.5–89.5% and 64.6–88.6% for MAE, Soxhlet and sonication, respectively. The precision of MAE was also better, ranging from 2.9–6.8%, as compared to 4.9–8.0% for soxhlet and 7.4–10.17% for sonication, based on three determinations.

9.7 RECOMMENDATIONS FOR MAE

Microwave-assisted extraction has been applied to a range of sample types (soils, sediments, water) for the environmental analysis of pollutants. Most applications described have utilised a sealed sample vessel, i.e. pressurised, for MAE. In this case, as safety is of primary importance in the analytical laboratory, it is unwise that a microwave system purchased for domestic use is used for extraction of pollutants with organic solvents. The work reported, using MAE, has commonly sought to investigate the dependency on operating parameters of the recovery of the analyte of interest. It is therefore possible to suggest some recommendations for the utilisation of pressurised MAE in the extraction of pollutants from solid matrices, e.g. soils.

Temperature: > 115 °C but < 145 °C
Pressure: Operating at < 200 psi
Microwave power: 100%
Extraction time (time at parameter): > 5 min but no need to extend beyond 20 min. The longest time is recommended if 12 vessels are to be extracted simultaneously.
Extraction solvent volume: 30–45 ml per 2–5 g of sample
Extraction solvent: hexane-acetone (1 : 1, v/v) has been most commonly used but other solvents also appear to be satisfactory on the basis of MAE (acetone and DCM).

These parameters are based on the use of a microwave system capable of delivering a minimum of 900 W of power. Ideally, the oven cavity should be equipped to allow ventilation of the cavity in the event of an organic vapour release. An additional safety feature, would be the inclusion of a solvent sensor that automatically shuts off the microwave source in the event of an organic solvent leakage, thus minimising the risk of fire. Extraction vessels should be mounted on a carousel arrangement that rotates through 360° allowing equal dissipation of the microwave energy to each sample vessel.

REFERENCES

1. E. Hasty and R. Revesz, *American Laboratory*, February (1995) 66.
2. D. Papoutsis, *Photonics Spectra*, March (1984) 53.
3. A. Abu-Samra, J.S. Morris and S.R. Koirtyohann, *Anal. Chem.*, **47** (1975) 1475.
4. K. Ganzler, A. Salgo and K. Valko, *J. Chromatogr.*, **371** (1986) 299.
5. V. Lopez-Avila, R. Young and W.F. Beckert, *Anal. Chem.*, **66** (1994) 1097.
6. V. Lopez-Avila, R. Young, J. Benedicto, P. Ho, R. Kim and W.F. Beckert, *Anal. Chem.*, **67** (1995) 2096.
7. I.J. Barnabas, J.R. Dean, I.A. Fowlis and S.P. Owen, *Analyst*, **120** (1995) 1897.
8. K.K. Chee, M.K. Wong and H.K. Lee, *J. Chromatogr.*, **723** (1996) 259.
9. F.I. Onuska and K.A. Terry, *Chromatographia*, **36** (1993) 191.
10. US Patent 5002784, J.R.J. Pare *et al.* 1991, Environment Canada.
11. V. Lopez-Avila, R. Young, R. Kim and W.F. Beckert, *J. Chromatogr. Sci.*, **33** (1995) 481.
12. J.R. Fish and R. Revesz, *LC-GC*, **14** (1996) 230.
13. K. Hummert, W. Vetter and B. Luckas, *Chromatographia*, **42** (1996) 300.
14. T.R. Steinheimer, *J. Agric. Food Chem.*, **41** (1993) 588.
15. S.J. Stout, A.R. daCunha and D.G. Allardice, *Anal. Chem.*, **68** (1996) 653.
16. M.P. Llompart, R.A. Lorenzo, R. Cela and J.R.J. Pare, *Analyst*, **122** (1997) 133.
17. V. Lopez-Avila, J. Benedicto, C. Charan, R. Young and W.F. Beckert, *Env. Sci. Technol.*, **29** (1995) 2709.
18. K.K. Chee, M.K. Wong and H.K. Lee, *Chromatographia*, **42** (1996) 378.
19. O.F.X. Donard, B. Lalere, F. Martin and R. Lobinski, *Anal. Chem.*, **67** (1995) 4250.
20. J. Szpunar, M. Ceulemans, V.O. Schmitt, F.C. Adams and R. Lobinski, *Anal. Chim. Acta*, **332** (1996) 225.
21. M.J. Vazquez, A.M. Carro, R.A. Lorenzo and R. Cela, *Anal. Chem.*, **69** (1997) 221.
22. F.I. Onuska and K.A. Terry, *J. Microcolumn Separations*, **7** (1995) 319.
23. F.I. Onuska and K.A. Terry, *J. High Resol. Chromatogr.*, **18** (1995) 417.
24. K.K. Chee, M.K. Wong and H.K. Lee, *Anal. Chim. Acta*, **330** (1996) 217.
25. J.R.J. Pare, J.M.R. Belanger, K. Li and S.S. Stafford, *J. Microcolumn Separations*, **7** (1995) 37.
26. V. Lopez-Avila, R. Young and N. Teplitsky, *J. AOAC Int.*, **79** (1996) 142.
27. J.R. Dean, I.J. Barnabas and I.A. Fowlis, *Anal. Proc.*, **32** (1995) 305.

APPENDIX A: SUPPLIERS OF COMMERCIAL MICROWAVE EXTRACTION SYSTEMS

CEM Corporation,
3100 Smith Farm Road,
P.O. Box 200,
Matthews,
NC 28106-0200,
USA

Milestone Corporation,
7289 Garden Road,
Suite 219,
Riviera Beach,
FL 33404,
USA

Prolabo Corporation,
24 Magnolia Court,
Lawrenceville,
NJ 08648,
USA

Questron Corporation,
P.O. Box 2387,
Princeton,
NJ 08543-2387,
USA

10

Accelerated Solvent Extraction

Accelerated solvent extraction (ASE) utilises organic solvents at high temperature and pressure to extract pollutants from environmental matrices. It was first proposed as a method (method 3545) in Update III of the USEPA SW-846 Methods, 1995.[1]

10.1 THEORETICAL CONSIDERATIONS

It has been suggested[2] that there are two main reasons why the use of organic solvents may give enhanced performance at elevated temperature and pressure compared to extractions at or near room temperature and atmospheric pressure: solubility and mass transfer effects, and disruption of surface equilibria.

10.1.1 SOLUBILITY AND MASS TRANSFER EFFECTS

Three factors are considered important:

- Higher temperature increases the capacity of solvents to solubilise analytes.
- Faster diffusion rates occur as a result of increased temperature.
- Improved mass transfer and hence increased extraction rates occur when fresh solvent is introduced, i.e. the concentration gradient is greater between the fresh solvent and the surface of the sample matrix.

10.1.2 DISRUPTION OF SURFACE EQUILIBRIA

As both temperature and pressure are important both are discussed separately.

Temperature effects

- Increased temperatures can disrupt the strong solute-matrix interactions caused by van der Waal's forces, hydrogen bonding, and dipole attractions of the solute molecules and active sites on the matrix.
- A decrease in viscosity of organic solvents is noted at higher temperature thus allowing improved penetration of the matrix, and hence improved extraction.

Pressure effects

- The utilisation of elevated pressures allows solvents to remain liquified above their boiling points.
- Extraction from within the matrix is possible, as the pressure allows the solvent to penetrate the sample matrix.

10.2 INSTRUMENTATION

Accelerated solvent extraction (ASE) was launched in the spring of 1995 by Dionex Corp., USA, as a fully automated sequential extraction system. A schematic diagram of a system is shown in Figure 10.1. The system (ASE 200) can operate with up to 24 sample-containing extraction vessels and up to 26 collection vials plus an additional 4 vial positions for rinse/waste collection. Three sample

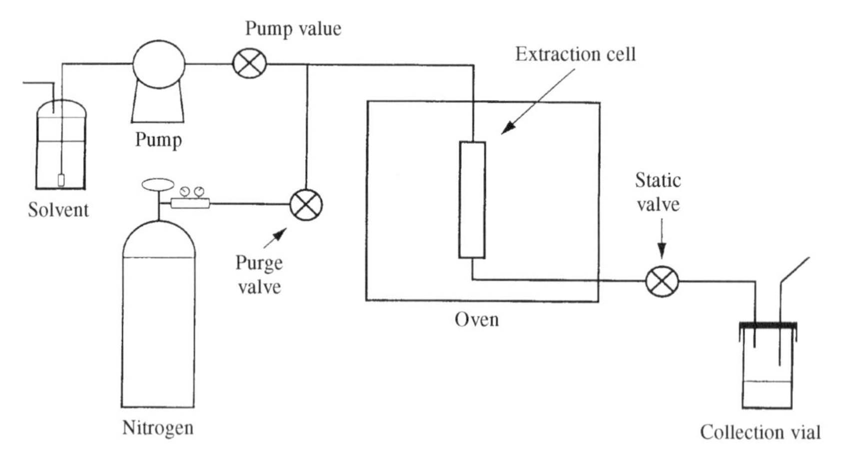

Figure 10.1 Schematic diagram of an accelerated solvent extraction system. Reproduced by permission of Dionex (UK) Ltd, Surrey

extraction vessel sizes are available: 11, 22 and 33 ml, with an internal diameter of 19 mm. The sample vessel has removable end caps that allow for ease of cleaning and sample filling. Each sample vessel is fitted with two finger-tight caps with compression seals for high pressure closure. To fill a sample vessel, one end cap is introduced and screwed on to finger-tightness. Then, a filter paper is introduced into the sample vessel, followed by the sample itself. Then, the other end cap is screwed on to finger-tightness and the sample vessel placed in the carousel. Through computer control the carousel introduces selected extraction vessels consecutively. An auto-seal actuator places the extraction vessel into the system and then returns the vessel to the carousel after extraction. The system is operated as follows: the sample cell, positioned vertically, is filled from top to bottom with the selected solvent or solvent mixture from a pump and then heated to a designated temperature (up to 200 °C) and pressure (up to 20 MPa or 3000 psi). These operating conditions are maintained for a prespecified time using static valves. After, the appropriate time (typically 5 min) the static valves are released and a few ml of clean solvent (or solvent mixture) is passed through the sample cell to exclude the existing solvent(s) and extracted analytes. This rinsing is enhanced by the passage of N_2 gas to purge both the sample cell and the stainless steel transfer lines. After gas purging, all extracted analytes and solvent(s) are passed through stainless steel tubing that punctures a septa (solvent resistant; coated with PTFE on the solvent side) located on top of the glass collection vials (40 or 60 ml capacity). If required, multiple extractions can be performed per extraction vessel. The arrival and level of solvent in the collection vial is monitored using an IR sensor. In the event of system failure, an automatic shut-off procedure is instigated.

10.3 APPLICATIONS

As the apparatus has not been commercially available for many years the literature is currently lacking any substantial input. However, some papers have appeared and these will be reviewed.

10.3.1 POLYCYCLIC AROMATIC HYDROCARBONS (PAHS)

Richter *et al.*[2] were able to demonstrate that PAHs could be extracted from (a) urban dust and (b) marine sediment using the following conditions: temperature, 100 °C; pressure, 2000 psi; extraction time, 5 min equilibriation plus 5 min static extraction; and extraction solvent, DCM : acetone (1 : 1, v/v). The results for the urban dust (SRM 1649) are shown in Figure 10.2 and for the marine sediment (HS-3) in Figure 10.3. In each case, excellent agreement was achievable compared to the certificate value (or noncertified value). In a more thorough study Dean[3] compared the results obtained from ASE with those obtained on the same soil type

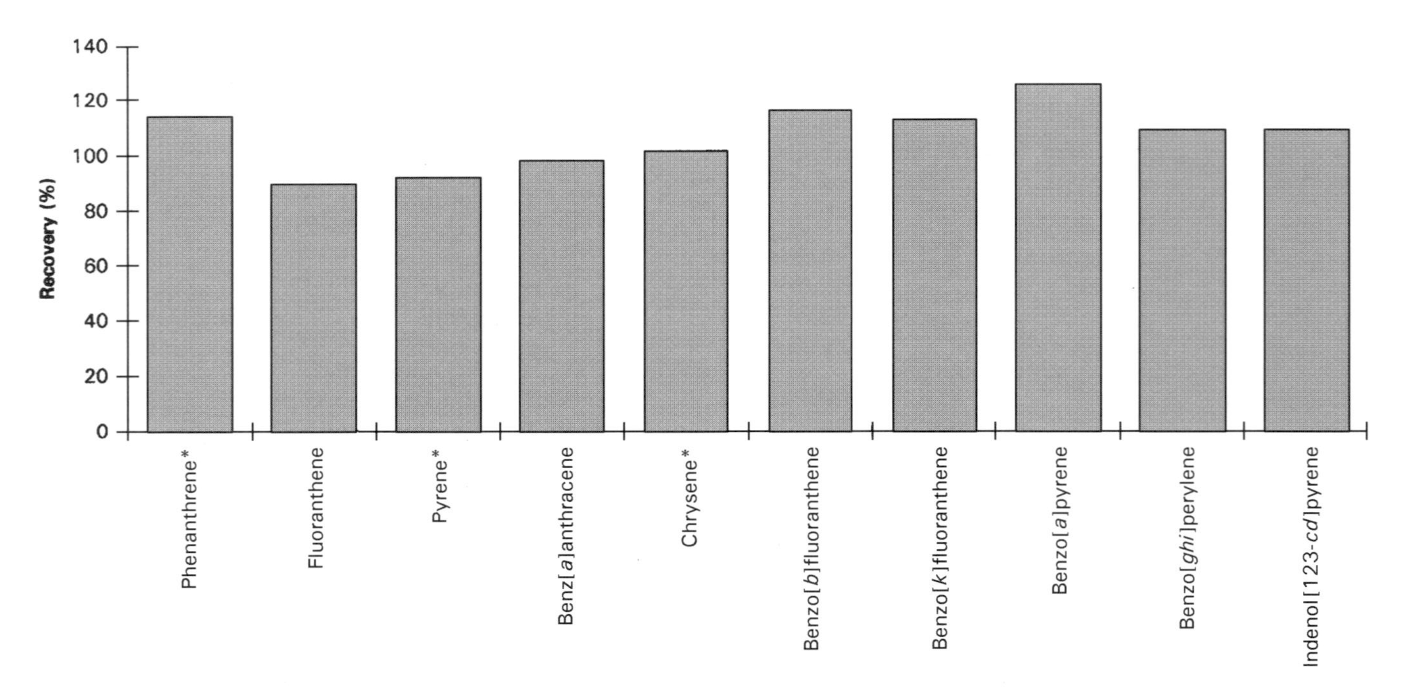

Figure 10.2 Percentage recovery of PAHs from urban dust (SRM 1649) (* denotes noncertified values). From Richter *et al.*, *Analytical Chemistry*, **68** (1996) 1033

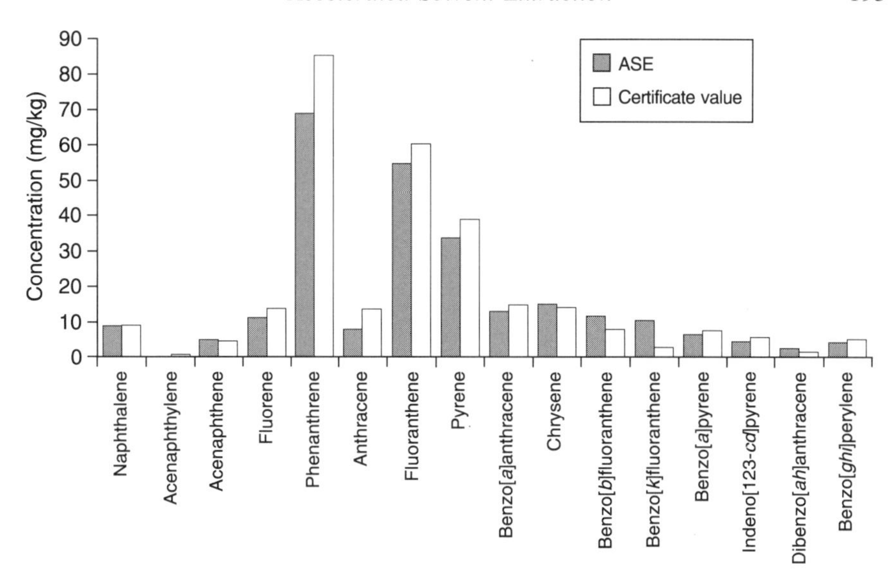

Figure 10.3 ASE of marine sediment (HS-3). From Richter *et al.*, *Analytical Chemistry*, **68** (1996) 1033

by SFE using methanol-modified CO_2, Soxhlet extraction (6 hours using DCM) and MAE (using acetone as the extraction solvent). The results using ASE (and the conditions described previously) compared favourably with those values obtained for the 16 individual PAHs by all the techniques studied, with the exception of Soxhlet extraction.

Recently, Saim *et al.*[4] evaluated the effect of ASE operating conditions on the recovery of PAHs from native, contaminated soil. The operating conditions evaluated were: pressure (1000–2400 psi), temperature (40–200 °C) and extraction time (2–16 min). The solvent mixture used for extraction was acetone : dichloromethane (1 : 1 v/v). Using an experimental design approach, based on a central composite design, each operating variable was evaluated. It was reported that for total PAHs recovered (the sum of 16 individual PAHs) no significance was found at the 95% confidence interval. However, when individual PAHs were evaluated it was found that some compounds had dependency upon the operating variables. The most significant operating variable was reported to be extraction temperature; naphthalene, chrysene and benzo[b]fluoranthene were found to have significance, at the 95% confidence interval, at an operating temperature of 40 °C. The effects of altering the solvent composition were also investigated. Using constant operating conditions (temperature, 100 °C; pressure, 14 MPa; and an extraction time of 5 min with 5 min equilibriation) the effects on recovery of solvent were investigated. The

Table 10.1 Extraction conditions for polycyclic aromatic hydrocarbons from contaminated soil[5]

Condition	Soxhlet extraction	Atmospheric microwave-assisted extraction	Pressurised microwave-assisted extraction	Supercritical fluid extraction	Accelerated solvent extraction
Sample size	10 g	2 g	2 g	2×1 g	7 g
Solvent	Dichloromethane, 150 ml	Dichloromethane (70 ml)	Acetone (40 ml)	CO_2 ($\sim$60 ml) + methanol ($\sim$12 ml)	Dichloromethane : Acetone, 1 : 1 (v/v) ($\sim$11 ml)
Temperature	Boiling point of solvent	Boiling point of solvent	120 °C	70 °C	100 °C
Pressure	Atmospheric	Atmospheric	Elevated pressure	250 kg cm^{-2} (approx. 2.45×10^7 Pa)	2000 psi (approx. 1.38×10^7 Pa)
Time	24 hours	20 min	20 min plus 30 min cooling	5 min static followed by 30 min dynamic	5 min equilibriation followed by 5 min static
Additional treatment	Evaporation to 10 ml	Evaporation to 5 ml	Evaporation to 5 ml	Evaporation to 5 ml	Evaporation to 5 ml

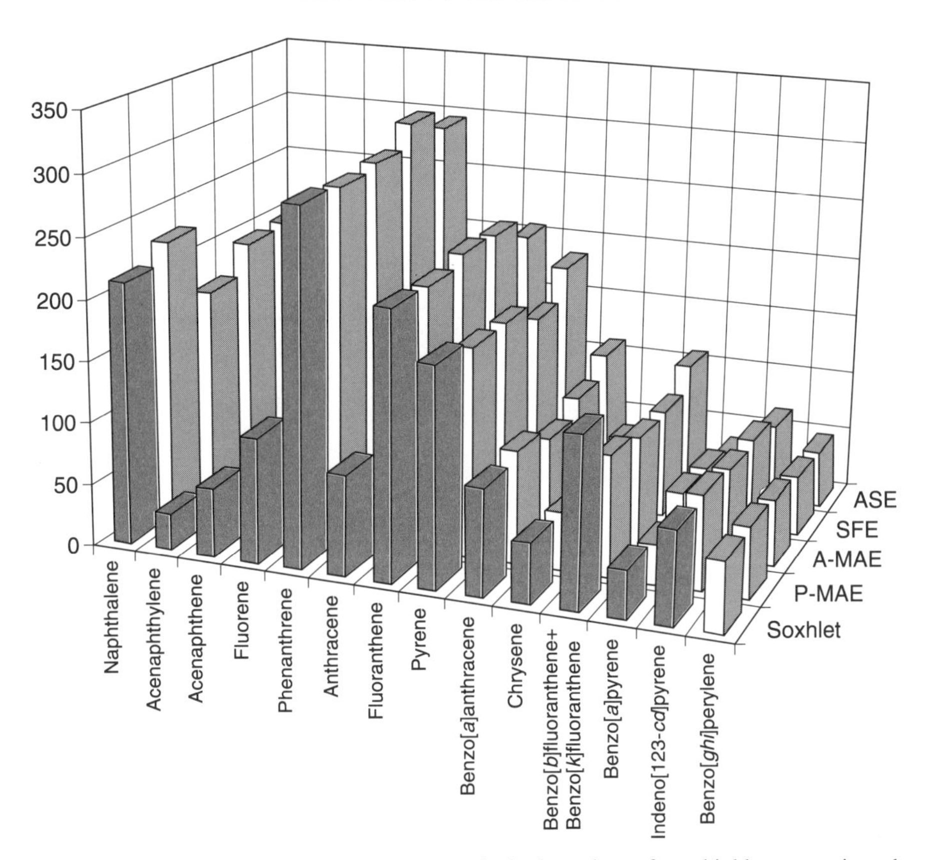

Figure 10.4 Extraction of polycyclic aromatic hydrocarbons from highly contaminated soil: comparison of Soxhlet extraction with pressurised MAE, atmospheric MAE, SFE and ASE. From N. Saim *et al.*, *Journal of Chromatography*, **791** (1997) 361

solvents evaluated, either as 1 : 1 combinations or as individual solvents, were as follows: DCM : acetone; acetone only; DCM only; methanol only; acetonitrile only; hexane : acetone; and hexane only. No dependence on recovery of individual PAHs, at the 95% confidence interval, was noted when solvent polarity was >1.89, i.e. all solvent systems except hexane. The same authors[5] evaluated the extraction of PAHs from contaminated soil using Soxhlet extraction, pressurised microwave-assisted extraction, atmospheric microwave-assisted extraction, supercritical fluid extraction and accelerated solvent extraction. Using the extraction conditions shown in Table 10.1, six subsamples of the soil were extracted by each technique. The results (Figure 10.4) indicate that the recovery of PAHs is dependent on the

Table 10.2 Extraction conditions for chimney and urban dust[6]

Condition	ASE[a]	Soxhlet
Sample size	4–10 g	4–10 g
Solvent	Toluene, 15 ml	Toluene, 250 ml
Temperature	180 °C	< 111 °C
Pressure	2000 psi	Atmospheric
Time	9 min heat up, 5 min static	18 hours
Analysis	GC-MS	GC-MS

[a] Two duplicate extractions were done on each sample. Results are the sum of both extractions.

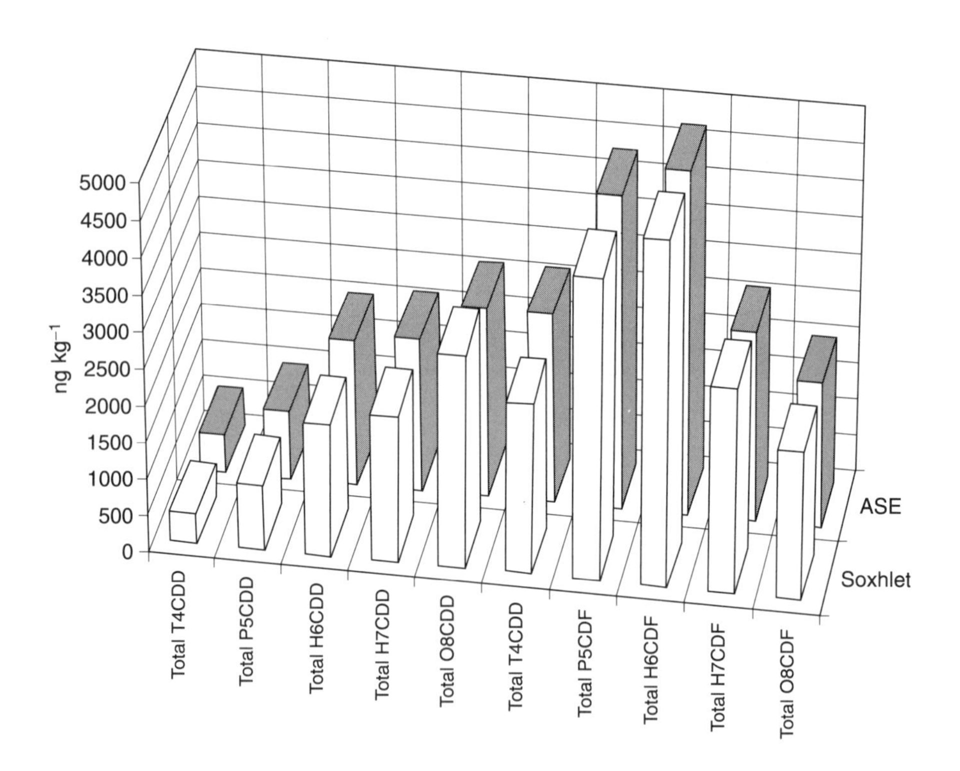

Figure 10.5 Extraction of total dioxins and furans from chimney brick using Soxhlet and ASE. From Richter *et al.*, *Chemosphere*, **34** (1997) 975

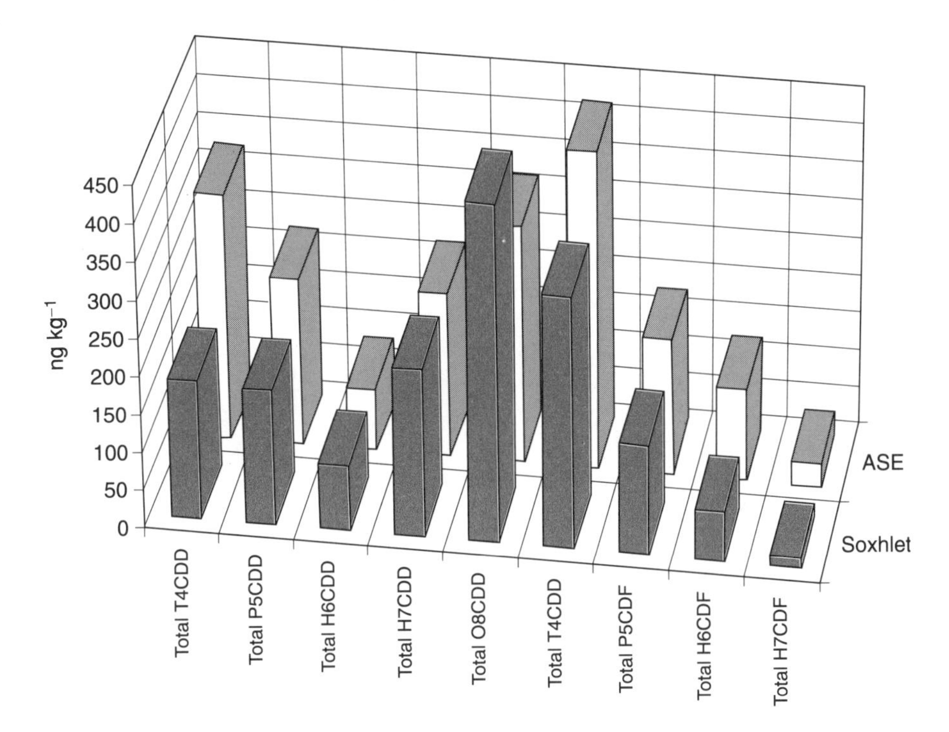

Figure 10.6 Extraction of total dioxins and furans from urban dust using Soxhlet and ASE. From Richter *et al.*, *Chemosphere*, **34** (1997) 975

extraction technique. The highest recoveries were consistently obtained by Soxhlet extraction.

10.3.2 POLYCHLORINATED BIPHENYLS (PCBS)

PCBs were extracted from both sewage sludge and oyster tissue with excellent recoveries.[2] The results were compared against those obtained using Soxhlet extraction with hexane as the extraction solvent and an extraction time of 6 hours. The results for the PCB congeners (28, 52, 101, 153, 138 and 180) ranged from 110 to 160% recovery for the sewage sludge and 86.3 to 90.0% recovery for the oyster tissue. The analyte concentration ranges were 160–200 ng g^{-1} for the sewage sludge and 50–150 ng g^{-1} for the oyster tissue.

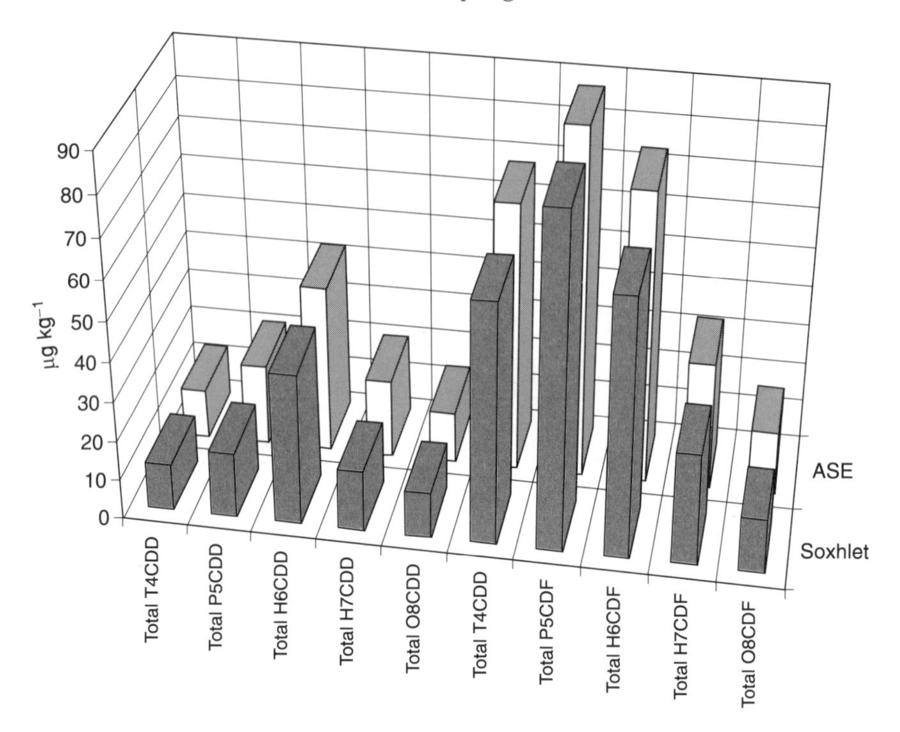

Figure 10.7 Extraction of total dioxins and furans from fly ash (1) using Soxhlet and ASE. From Richter *et al.*, *Chemosphere*, **34** (1997) 975

10.3.3 DIOXINS AND FURANS

Soxhlet extraction has been compared with ASE for the extraction of polychlorinated dibenzo-*p*-dioxins and polychlorinated dibenzofurans from soil, sediment, chimney brick, urban dust and fly ash.[6] The extraction conditions for chimney brick and urban dust by both ASE and Soxhlet are shown in Table 10.2. The fly ash sample was initially treated with either 6 M HCl for 30 min and then rinsed thoroughly with distilled water or 5% (v/v) glacial acetic acid was added to the toluene for the ASE extraction only. For the soil and sediment samples, ASE was done using a double static extraction period (2 × 5 min), instead of combining the extracts from two separate extractions. All other extraction conditions are identical to those described in Table 10.2. Additional clean-up of the samples was done according to the German draft method VDI 3499 for chimney brick, urban dust and fly ash samples; the clean-up for the sediment and soil samples has been described elsewhere.[7] The results, expressed as group totals (congener values

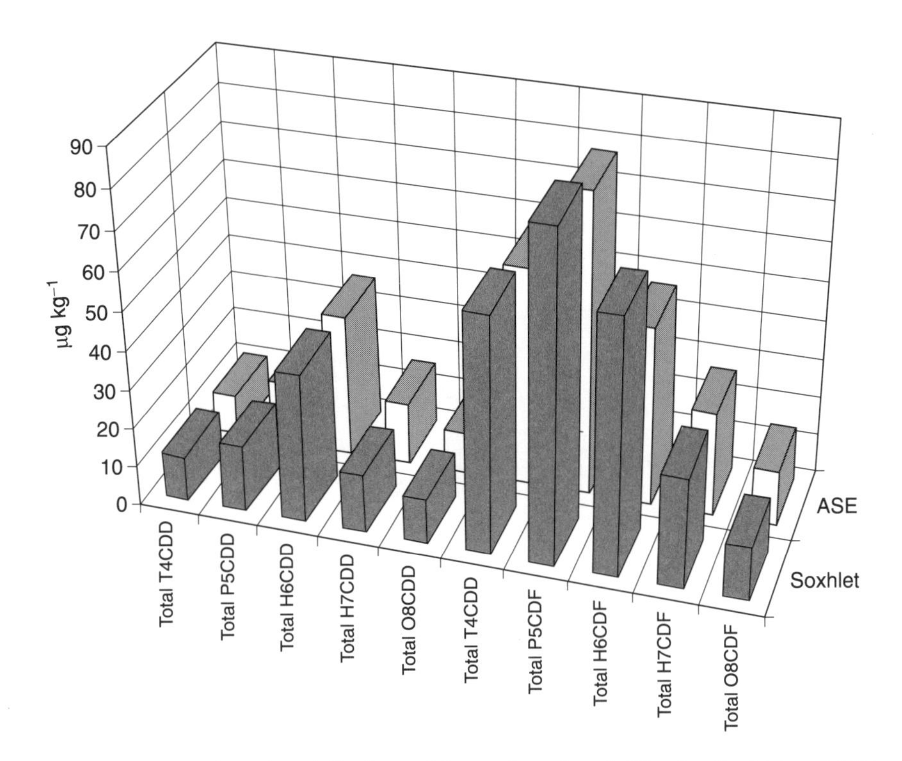

Figure 10.8 Extraction of total dioxins and furans from fly ash (2) using Soxhlet and ASE. From Richter *et al.*, *Chemosphere*, **34** (1997) 975

available in the paper), are shown in Figures 10.5–10.12. It is clear that ASE compares favourably with Soxhlet extraction for the extraction of polychlorinated dibenzo-p-dioxins and dibenzofurans from a range of environmental samples. The lower organic solvent consumption (5 ml for ASE against 250 ml for Soxhlet) and faster extraction time (<25 min for ASE compared to 18 hours for Soxhlet) demonstrates that ASE is an efficient alternative method of extraction to classical Soxhlet extraction.

10.3.4 PESTICIDES

As part of an equivalence study Ezzell *et al.*[8] investigated the extraction of 32 organophosphorus pesticides (OPPs) from three soil matrices (clay, loam and sand). The soils were spiked at two different levels, i.e. in the range 53.5–500 $\mu g\ kg^{-1}$ and 535–5000 $\mu g\ kg^{-1}$, depending upon the OPP. The results indicated that comparable

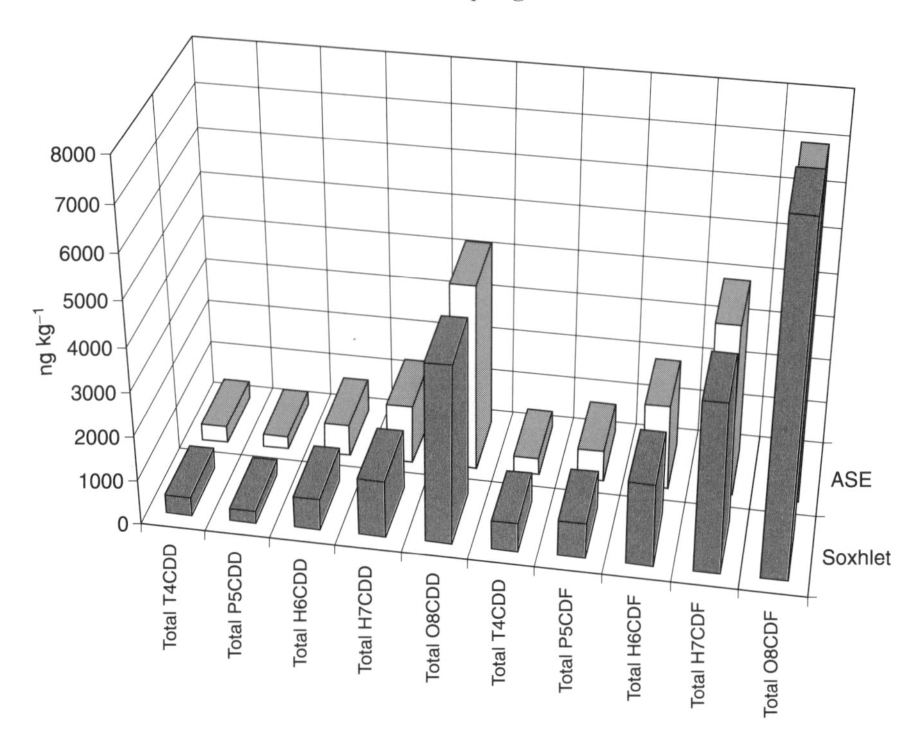

Figure 10.9 Extraction of total dioxins and furans from EC-2 using Soxhlet and ASE. From Richter *et al.*, *Chemosphere*, **34** (1997) 975

results were obtainable with an 18-hour DCM Soxhlet extraction and were independent of the soil matrix.

In a related area Obana *et al.*[9] utilised ASE to extract 19 OPPs from spiked foods. The OPPs were extracted at 100 °C, pressure 1500 psi in under 20 min. Wet samples were extracted after mixing the food samples with Extrelut drying agent. The foodstuffs (flour, grapefruit juice, orange juice, broccoli) were spiked at levels between 0.05 and 0.1 ppm and their recoveries assessed. Initial work evaluated the effect of extraction solvent on the extraction recovery of 17 OPPs (0.05 ppm) from flour (10 g). The solvents evaluated were: cyclohexane–acetone, dichloromethane–acetone and ethylacetate–acetone (all in 1 : 1 ratios). The recoveries of most of the pesticides were in the range 83–115% with RSDs ($n = 3$) of $<10\%$ using cyclohexane–acetone. With ethylacetate–acetone the RSDs ($n = 3$) varied between 19 and 34%. Dichlorvos was only poorly recovered in all three solvent systems (approx. 42%). It was inferred that dichlorvos was lost during the sample preparation due to its volatility. Peculiar results were obtained for dimethoate in

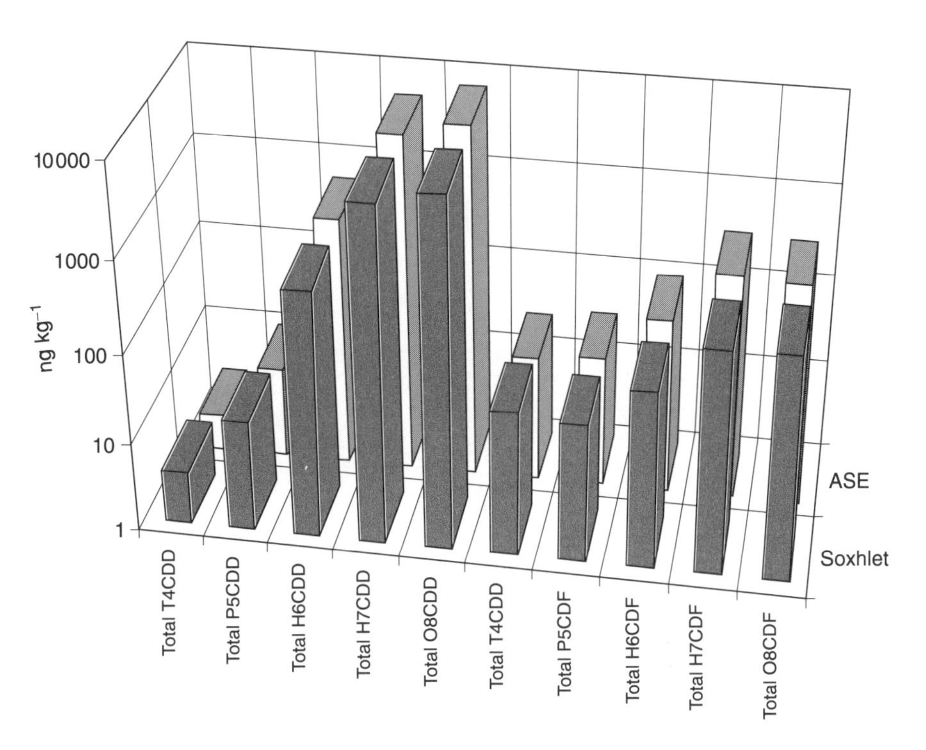

Figure 10.10 Extraction of total dioxins and furans from HS-2 using Soxhlet and ASE. From Richter *et al.*, *Chemosphere*, **34** (1997) 975

each solvent system with recoveries ranging from 49.1% in cyclohexane–acetone to 199% in ethylacetate–acetone. No explanation was offered in the paper. Using cyclohexane–acetone 19 OPPs were extracted from both orange and grapefruit juice. The average recoveries ($n = 3$) ranged from 82–105% in grapefruit juice to 51–110% in orange juice, with the exception of methamidophos and acephate which were only poorly recovered ($< 11\%$). The precision of the results was poor in most cases. An attempt was made to improve the recovery of methamidophos and acephate in orange juice by adjusting the ASE conditions (temperature, 50, 100 and 150 °C; pressure, 1500, 2000 and 2500 psi; and extraction time, 5, 10 and 15 min) and extraction solvent (cyclohexane–acetone, toluene–acetone, dichloromethane–acetone, ethylacetate–acetone, acetonitrile and ethylacetate). No significant enhancement in recovery was noted by altering the ASE conditions. However, by using ethylacetate approx. 50% recovery was achievable. Therefore using ethylacetate as the extraction solvent 19 OPPs were extracted from orange juice (spike level 0.1 ppm), grapefruit juice (spike level 0.1 ppm), broccoli (spike

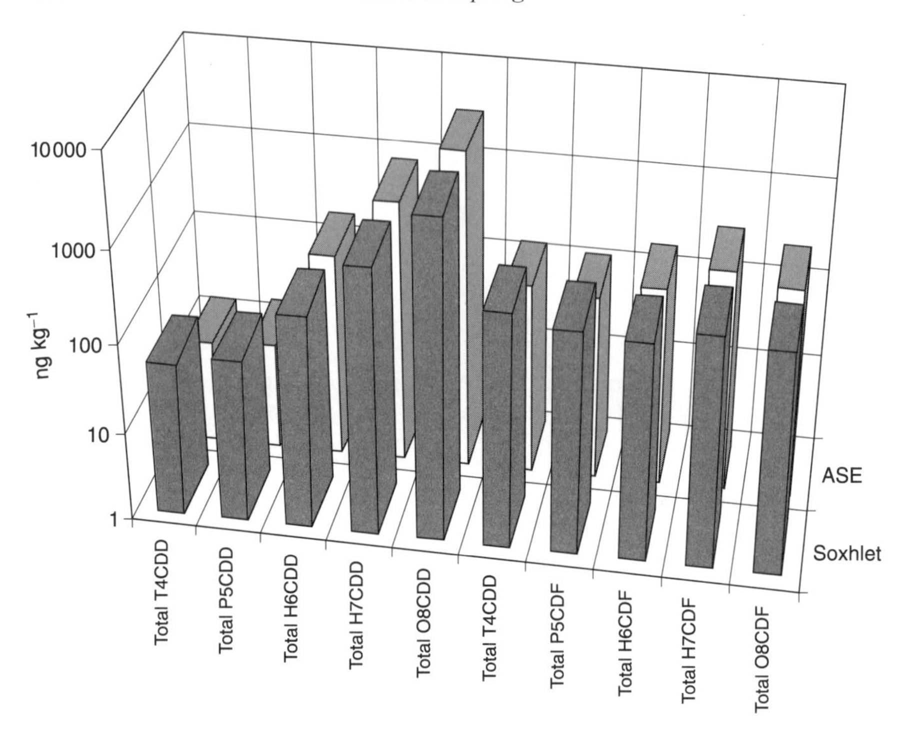

Figure 10.11 Extraction of total dioxins and furans from Hamilton Harbour sediment using Soxhlet and ASE. From Richter *et al.*, *Chemosphere*, **34** (1997) 975

level 0.1 ppm) and flour (spike level 0.05 ppm). The results are shown in Figure 10.13. The average recoveries were 79.3, 87.4, 90.4 and 82.9%, respectively. The technique was compared with a hexane extraction for the analysis of food samples which had been found to be contaminated with pesticides. The results for the three OPPs (chlorpyrifos, phosalone and malathion) in banana, okra and sweetie were all lower by ASE, except malathion in sweetie which was in agreement. However, the precision by ASE was always better than the hexane extraction. The workers suggest that the use of an automated ASE system offers potential advantages over existing methods for pesticide residue analysis.

The extraction of hexaconazole, a broad spectrum systemic triazole fungicide, from weathered soils has been evaluated.[10] Soxhlet extraction was compared with microwave-assisted extraction, supercritical fluid extraction and accelerated solvent extraction for the extraction of hexaconazole from Canadian soil. The soil had been previously weathered over periods of time varying from 0 to 52 weeks, air dried for 48 hours and deep frozen prior to transportation to the UK. Two soil types were

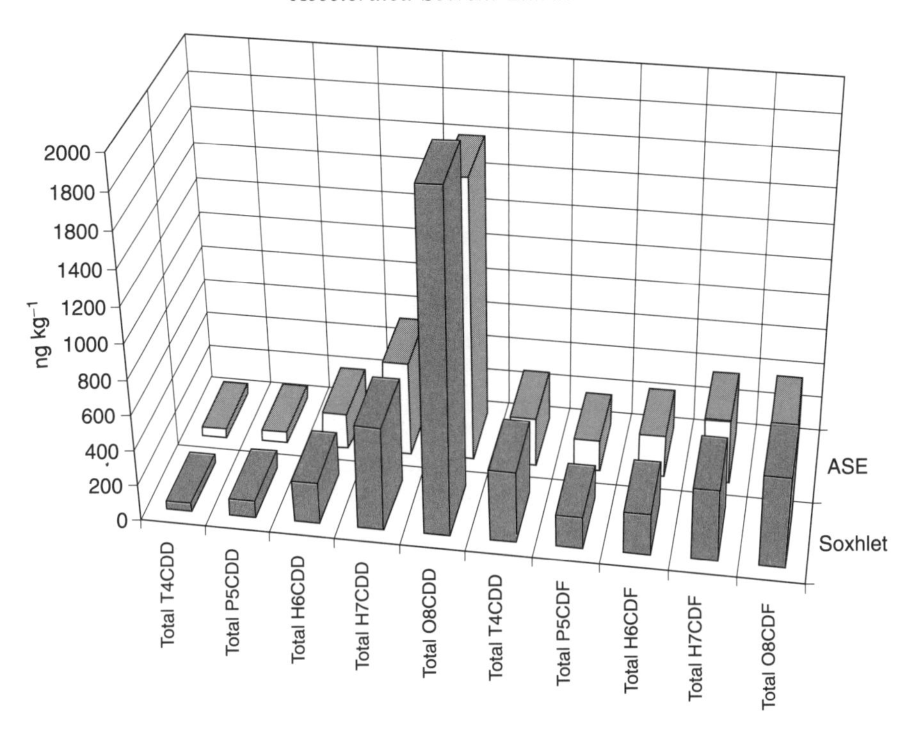

Figure 10.12 Extraction of total dioxins and furans from Parrots Bay sediment using Soxhlet and ASE. From Richter *et al.*, *Chemosphere*, **34** (1997) 975

extracted: a sandy loam soil (organic matter 5.7%); and a sandy clay soil (organic matter, 1.5%). The extraction conditions are described in Table 10.3. The results, for the sandy clay soil (organic matter, 1.5%) (Figure 10.14), indicate that comparable recoveries can be achieved by all three extraction methods, as compared to Soxhlet extraction. This was not the case with the sandy loam soil (organic matter, 5.7%) where it is observed (Figure 10.15) that the recoveries by SFE and MAE were approximately only 50% of those obtained by Soxhlet extraction. In contrast to these results were those obtained by ASE, which compared favourably with those obtained by Soxhlet extraction (Figure 10.14). The best precision in all cases was obtained using ASE. This is probably due to the use of an automated system.

10.3.5 PHENOLS

A home-made ASE system has been applied to the extraction of seven phenols from slurry spiked soils.[11] In this work, a preliminary investigation into the factors

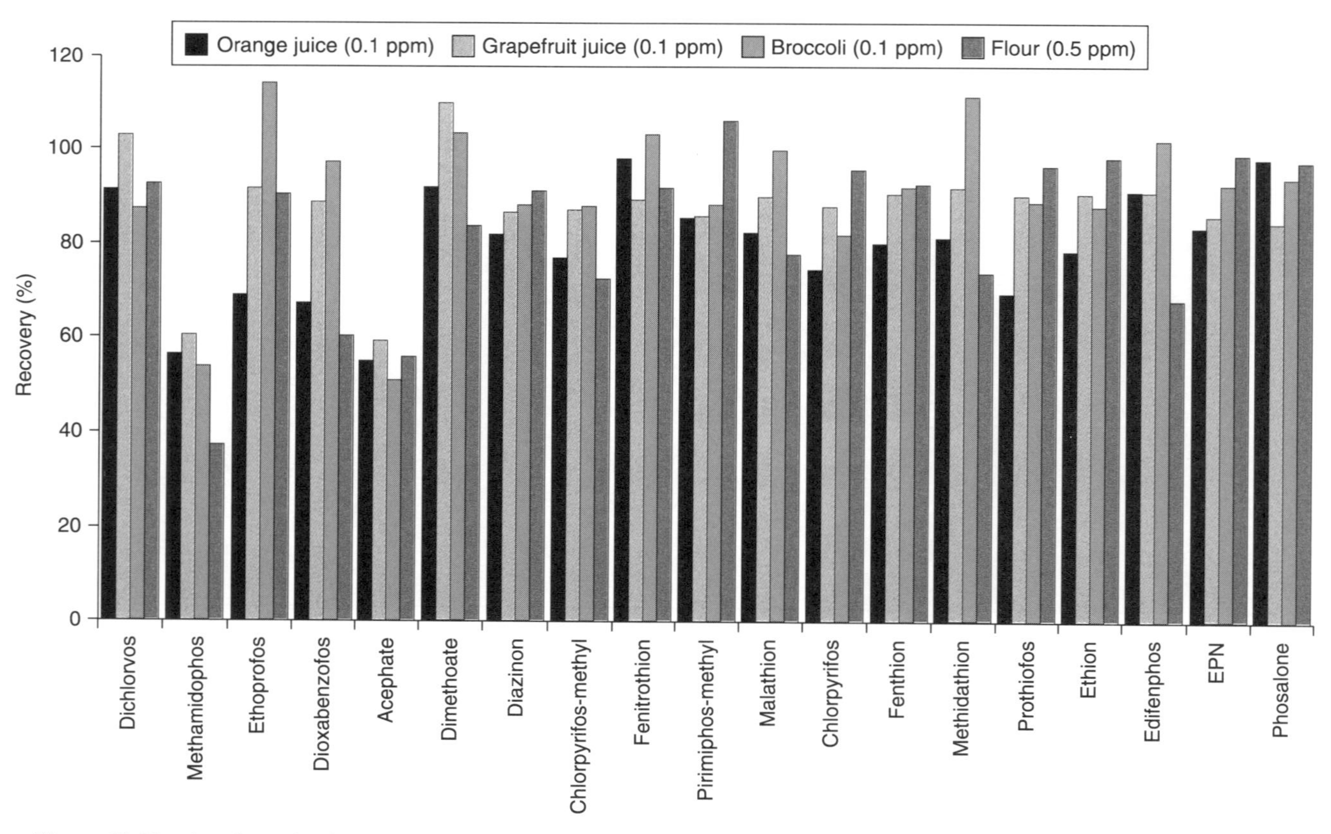

Figure 10.13 Accelerated solvent extraction of OPPs from spiked foodstuffs using ethyl acetate. From Obana *et al.*, *Analyst*, **122** (1997) 217

Table 10.3 Extractions conditions for hexaconazole from weathered Canadian soil[10]

Condition	Soxhlet extraction	Microwave-assisted extraction	Supercritical fluid extraction	Accelerated solvent extraction
Sample size	40 g	5 g	4 g	5 g
Solvent	Acetonitrile : Water (1 : 1), 80 ml	Acetone (30 ml)	CO_2 (∼40 ml) + methanol (∼8 ml)	Acetone (∼11 ml)
Temperature	–	115 °C	55 °C	100 °C
Pressure	Atmospheric	Elevated pressure	250 kg cm^{-2} (approx. 2.45 × 10^7 Pa)	2000 psi (approx. 1.38 × 10^7 Pa)
Time	6 hours	15 min plus 30 min cooling	5 min static followed by 20 min dynamic	10 min
Additional treatment	Liquid–liquid separation with 5 ml of dichloromethane; reduced to dryness and then rediluted with 1 ml of dichloromethane. Further clean-up using a silica solid phase extraction cartridge	Clean-up using a C18 solid phase extraction cartridge. All samples filtered and reduced in volume to 1 ml prior to analysis	Clean-up using a C18 solid phase extraction cartridge (sandy loam soil only). All samples filtered and reduced in volume to 1 ml prior to analysis	All samples filtered and reduced in volume to 1 ml prior to analysis

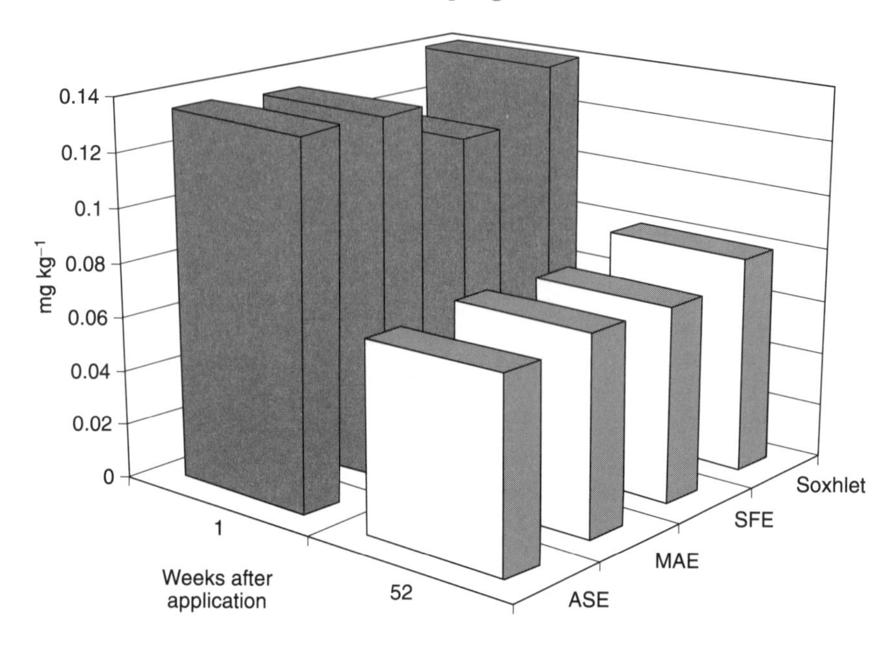

Figure 10.14 Hexaconazole from weathered Canadian soil (sandy clay soil). From Frost *et al.*, *Analyst*, **122** (1997) 895

that might influence the ASE of the spiked phenols was undertaken. As the system was limited to certain operating constraints, i.e. a maximum temperature of 70 °C, its direct relevance to that of the commercial system is limited. Nevertheless the influence of the three main operating variables, i.e. temperature, pressure and extraction time were investigated using acetonitrile as the extraction solvent. The limits of these variables were: temperature, 30–70 °C; pressure, 4–20 MPa; and extraction time, 5–25 min. The results of the central composite design indicated that only one phenol (2-methylphenol) was influenced by the operating variables, i.e. pressure and extraction time. It was also important to note that in every case, the recovery of 2,4-dimethylphenol was poor (mean recovery 24.5%). (Note: In the case of shake-flask extraction 2,4-dimethylphenol was not recovered at all.) Another interesting feature was observed when comparing results with those obtained using a shake-flask approach. The shake-flask method was done using an acetonitrile-water mixture whereas the ASE was done using acetonitrile only. This resulted in the chromatograms shown in Figure 10.16. It is clearly observed that the use of acetonitrile only (hot solvent extraction) resulted in cleaner chromatograms than those obtained when using an acetonitrile–water mixture.

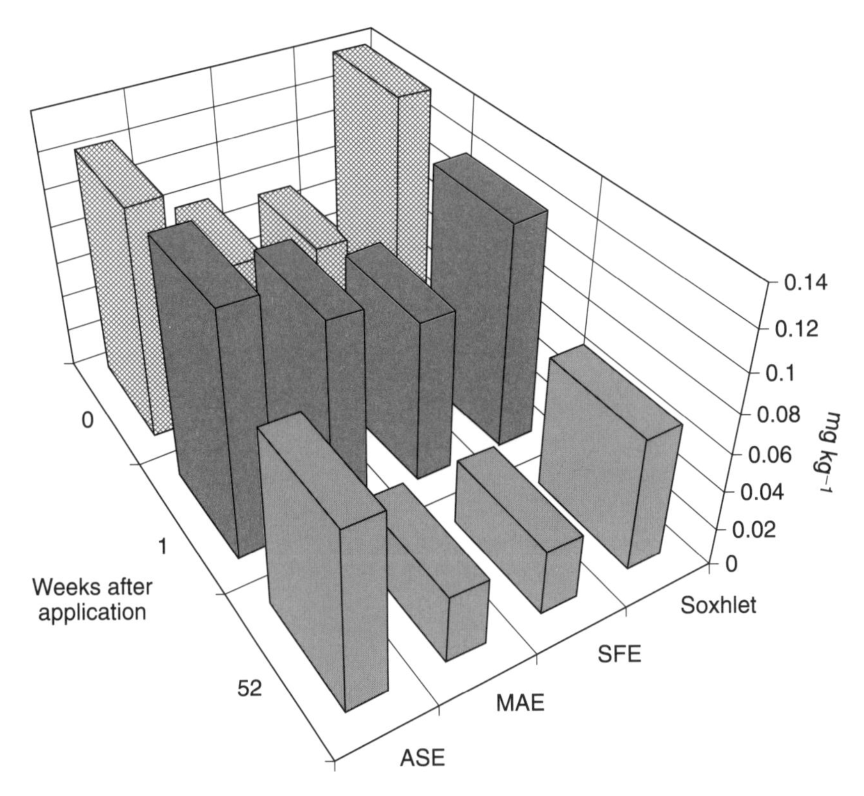

Figure 10.15 Hexaconazole from weathered Canadian soil (sandy loam soil). From Frost *et al.*, *Analyst*, **122** (1997) 895

10.4 RECOMMENDATIONS FOR ASE

As most of the work published has focused on the manufacturer's recommended operating conditions and the EPA Method[1] it is sufficient to reiterate those conditions for use.

Temperature: 100 °C
Pressure: 2000 psi
Extraction time: 5 min equilibriation plus 5 min static extraction
Extraction solvent: DCM : acetone (1 : 1, v/v)

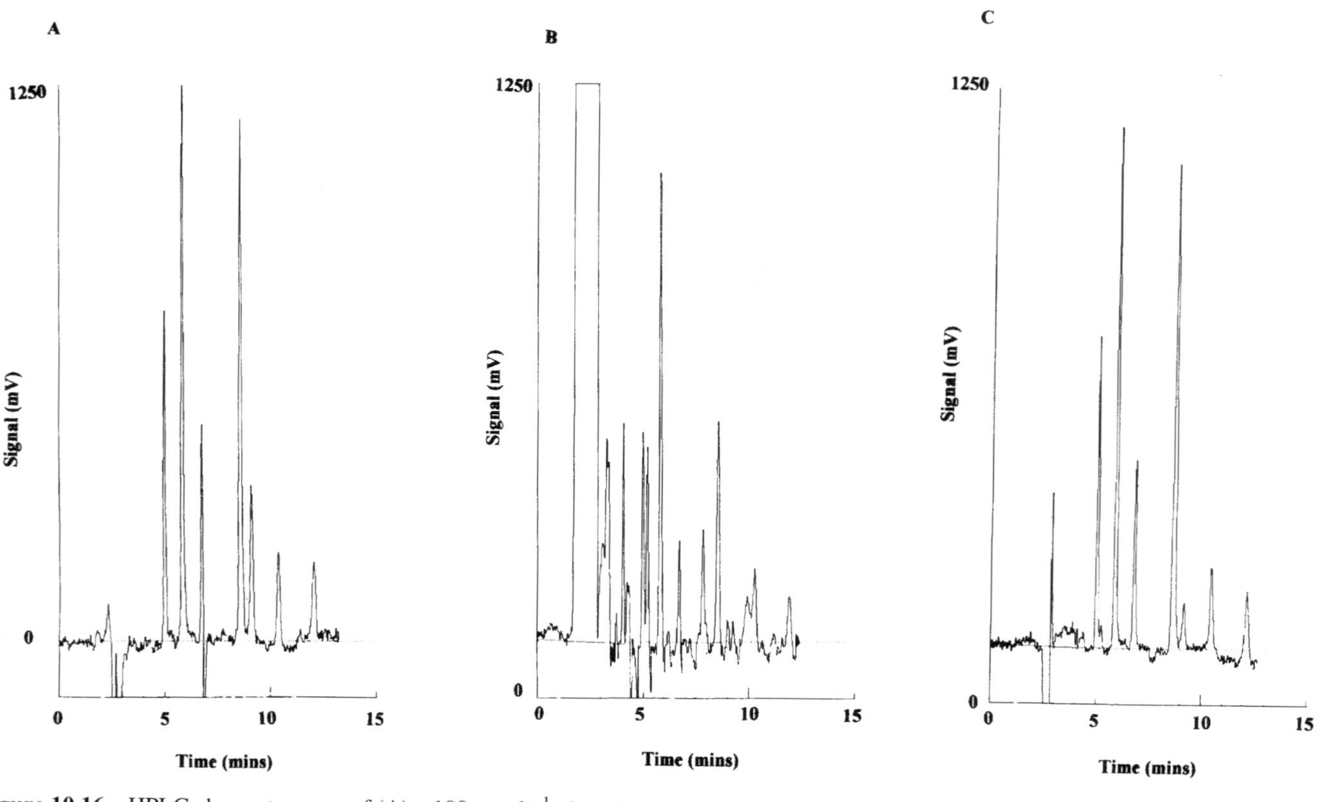

Figure 10.16 HPLC chromatograms of (A) a 100 ng ml^{-1} phenol standard mixture; (B) shake-flask extraction; and (C) accelerated solvent extraction. From Dean *et al.*, *Analytical Communications*, **33** (1996) 413

REFERENCES

1. *Testing methods for evaluating solid waste, Method 3545*, USEPA SW-846, 3rd edition, Update III, Washington, DC, July 1995.
2. B.E. Richter, B.A. Jones, J.L. Ezzell, N.L. Porter, N. Avdalovic and C. Pohl, *Anal. Chem.*, **68** (1996) 1033.
3. J.R. Dean, *Anal. Comm.*, **33** (1996) 191.
4. N. Saim, J.R. Dean, Md. P. Abdullah and Z. Zakaria, *Anal. Chem.*, **70** (1998) 420.
5. N. Saim, J.R. Dean, Md. P. Abdullah and Z. Zakaria, *J. Chromatogr.*, **791** (1997) 361.
6. B.E. Richter, J.L. Ezzell, D.E. Knowles, F. Hoefler, A.K.R. Mattulat, M. Scheutwinkel, D.S. Waddell, T. Gobran and V. Khurana, *Chemosphere*, **34** (1997) 975.
7. Ontario Ministry of the Environment and Energy, Method E3151B (1993): *The determination of polychlorinated dibenzo-p-dioxins and dibenzofurans in soil and sediment by GC-MS.*
8. J.L. Ezzell, B.E. Richter, W.D. Felix, S.R. Black and J.E. Meikle, *LC-GC*, **13** (1995) 390.
9. H. Obana, K. Kikuchi, M. Okihashi and S. Hori, *Analyst*, **122** (1997) 217.
10. S.P. Frost, J.R. Dean, K.P. Evans, K. Harradine, C. Cary and M.H.I. Comber, *Analyst*, **122** (1997) 895.
11. J.R. Dean, A. Santamaria-Rekondo and E. Ludkin, *Anal. Comm.*, **33** (1996) 413.

11

Comparison of Extraction Methods

The requirements of any extraction technique are that it can produce valid data, rapidly, with minimum operator involvement, be cost effective and satisfy safety considerations for both the operator, other personnel and its location within the operating environment. Tables 11.1 and 11.2 compare the merits of liquid and solid extraction techniques, respectively.

11.1 FUTURE DEVELOPMENTS IN SAMPLE PREPARATION

Crystal-ball gazing is always difficult in any subject area, and this is no exception. However, some newer methods of sample preparation have already been included in this book. For liquid samples, the use of solid phase microextraction (SPME) is sure to have a bigger impact in the next decade. The application of flow technology for the online separation/extraction of analytes, e.g. online SPE-GC/LC;[1–3] membrane extraction with a sorbent interface (MESI);[4] and liquid–liquid extraction in continuous flow analysis[5] are other interesting developments. While for solid samples, the use of accelerated solvent extraction (ASE) has made a large initial impact due to its acceptance as an EPA method. Competition from microwave-assisted extraction is sure to drive each manufacturer to improve their respective technology further.

If we are looking forward to the next millennium it is important to consider what might be required in the laboratory of the future. In terms of solvent usage any attempts to minimise solvent consumption and hence its subsequent disposal will be required. The introduction of more automated systems of analysis for routine samples will also be beneficial. The continued expansion of the use of automated systems (autosampler-type devices and robotics) can only be useful as far as laboratory productivity is concerned. That is not to say that the trained

Table 11.1 Comparison of extraction techniques: liquids

	Liquid–liquid	Purge and trap	SPE	SPME
Description of system	Sample is partitioned between two immiscible solvents; continuous and discontinuous operation possible	Aqueous sample is purged with a gas followed by trapping on a suitable adsorbent. Analytes desorbed using heat and transferred direct to GC. An on-line system of analysis	Analyte retained on a solid adsorbent; extraneous sample material washed from sorbent. Desorption of analyte using organic solvent	Analyte retained on a sorbent-containing fibre attached to a silica support. Fibre protected by a syringe barrel when not in use. Most commonly found for GC applications; recently become available for HPLC
Acceptability	Wide acceptance for isolating organic compounds	Commonly used for volatile organic compounds	Widely accepted	Gaining in popularity; new technology
Extraction time	Discontinuous: 20 min and continuous: up to 24 hours	10–20 min	10–20 min	10–60 min (requires optimisation)
Solvent usage	3 × 60 ml for discontinuous; up to 500 ml for continuous	No solvent usage	Organic solvent required for wetting sorbent and elution of analyte (10–20 ml)	No solvent usage required
Cost	Low cost	Moderate cost for equipment	Relatively low cost (use of an SPE manifold). Cartridges are disposable	Relatively low cost (replacement fibre and syringe barrel holders); available as an automated system

Ease of operation	Easy	Automated systems available	Method development required; automated and robotic systems available for routine operation	Relatively easy to use; care required because of fragility of fibre. Automated systems available
Sample size	1 litre	5 to 25 ml	Various (1 ml to 1 litre)	Various (1 ml to 1 litre)
Approval of methods	EPA methods (method 3510 and 3520)	EPA method (method 5030)	EPA method (method 3535)	None at present
Main disadvantages	Concentration of sample required after extraction	Applicable to volatile organic compounds only; selection of adsorbent required. Optimisation of system required	Method development required; choice of sorbent and optimisation of sorbent selectivity	Method development required; choice of sorbent and optimisation of procedure required
Suppliers of equipment	Commonly available	Commonly available from suppliers of GC equipment	Commonly available (large choice of sorbent types available)	Only available from selected suppliers (Supelco, Varian)

Table 11.2 Comparison of extraction techniques: solids

	Soxhlet	Soxtec	Sonication	Supercritical fluid extraction (SFE)	Microwave-assisted extraction (MAE)	Accelerated solvent extraction (ASE)
Description of system	Utilises cooled condensed solvents to pass over the sample contained in an extraction thimble to extract the analytes. Uses specialised glassware and heating apparatus	Sometimes called automated Soxhlet. Soxtec places the sample into the boiling solvent and then flushes clean solvent over the sample. Faster than Soxhlet	Sample is covered with organic solvent, then a sonic horn is placed inside the beaker with the solvent and sample	SFE uses supercritical CO_2 with and without an organic modifier to extract analytes; pressures of up to 680 atm and temperatures up to 150 °C are used. Collection is either into a liquid solvent or onto a solid phase trap. The latter requires desorption with an organic solvent rinse	Microwaves used to heat a sample at either atmospheric pressure and near room temperature (open system) or at elevated pressures ($<$ 200 psi) and high temperatures ($>$ boiling point of the solvent e.g. DCM is 120 °C) (closed system)	Sample is extracted under high pressure (2000 psi) and temperature (100 °C). Solvent and analytes are flushed from the extraction vessel using a small volume of fresh solvent and a N_2 purge. Fully automated
Acceptability	Used as the bench-mark extraction technique by which others are judged	Considered to be the automated version of Soxhlet	Considered to be less reliable than Soxhlet	Considered to be difficult to develop reliable methods	Not been commercially available for organic extraction until the 1990s	Later on the market than microwaves
Extraction time	4–24 hours commonly used	Reduced time compared to Soxhlet, i.e. 4–5 hours	Initiated for periods of 3–15 min. Typically three time periods might be used	Relatively short extraction times (30–60 min)	Rapid extraction times (approximates to 5 min per sample vessel–up to 12 sample vessels can be used i.e. 1 hour)	Rapid extraction times (15 min)

Solvent usage	250–500 ml per extraction	40–50 ml per extraction	150–300 ml per extraction	Minimal solvent usage (10–30 ml)	Moderate solvent usage (40 ml)	Minimal solvent usage (15–25 ml)
Cost	Very inexpensive	Inexpensive	Relatively inexpensive	High cost	Moderate cost	High cost
Ease of operation	Easy to use	Easy to use	Relatively easy to use but labour intensive	Considered to be difficult to operate	Easy to use because of acceptance of domestic ovens	Easy to use
Sample size	> 10 g	> 10 g	up to 5 g	> 1 g	up to 5 g	> 10 g possible
Sequential or simultaneous extraction	Individual units required for multiple sample extraction	Systems available for 2 or 6 samples simultaneously	Sequential operation	Simultaneous or sequential systems available	Simultaneous operation of up to 12 samples	Sequential operation of up to 24 samples
Approval of methods	EPA	EPA	EPA	EPA	None for organic extraction	EPA
Main disadvantages	Large solvent usage	Solvent usage; more expensive than Soxhlet	Not automatable; labour intensive	Matrix effects identified; high capital cost	Requires polar solvents; filtration of sample required after cooling	High capital cost; new technique
Some suppliers of equipment		Tecator		Isco, HP, Dionex, Suprex, JASCO	CEM, Prolabo, Milestone	Dionex

analytical scientist will have no job! Rather that he/she can spend more time developing and investigating faster and/or better methods of analysis.

The advent of laboratories on a 'chip' is currently in its early stages of development. While the concurrent development of chromatograms on a chip presents an interesting development in nanotechnology it will be more interesting to see the novel approaches that will be required to prepare and then introduce real samples into them. Developments in this direction are currently under way through the work on capillary electrophoresis (CE) and capillary electrochromatography (CEC).

Whatever developments do occur, in terms of scientific instrumentation for sample preparation, will require innovation followed by extensive development and application prior to acceptance and replacement of existing technology. This will require the involvement of engineers and analytical scientists to collaborate on new projects. It is perhaps sobering to think that as you gaze around your laboratory at your newest GC or HPLC system that the sample preparation techniques that you are probably using have been around for most of this century (or last)!

REFERENCES

1. V. Pichon and M.C. Hennion, *J. Chromatogr.*, **665** (1994) 269.
2. U.A.Th. Brinkman and R.J.J. Vreuls, *LC-GC Int.*, **8** (1995) 694.
3. S. Ollers, M. van Lieshout, H.G. Janssen and C.A. Cramers, *LC-GC Int.*, **10** (1997) 435.
4. M.J. Yang, M. Adams and J. Pawliszyn, *Anal. Chem.*, **68** (1996) 2782.
5. M. Valcarel, 'Liquid–liquid extraction in continuous flow analysis' in *Developments in Solvent Extraction*, S. Algret (ed.), Ellis Horwood, Chichester (1988).

General Index

Chemical Index